'*Introducing Nonroutine Math Problems* provides a tested, creative, and practical approach to developing mastery in problem solving through mathematics – but what is more revelatory is that it cultivates the deep processes that are equally relevant to the immediate problems of our lives and our world. For both self and for schooling, it is transformative, masterful, and meaningful.'
  – **Tobin Hart, Ph.D.**, *Professor of Psychology, University of West Georgia*

'Bob London's research-based, holistic approach is discussed with depth and clarity in *Introducing Nonroutine Math Problems to Secondary Learners*. It is supported by an extensive use of strategies and examples, thereby making it an excellent context and resource for exploring the CCMP.'
  – **Dr Marian de Souza**, *Retired from Australian Catholic University, Australia*

'*Introducing Nonroutine Math Problems* is an amazing contribution to mathematics education and to learning that is relevant, holistic, and engaging. Written in a personal style, Dr. London guides educators through each process and activity. This book will change your classroom and touch your students' lives.'
  – **Sam Crowell**, *co-author*, Emergent Teaching

Introducing Nonroutine Math Problems to Secondary Learners

Offering secondary math educators an innovative holistic and process-orientated approach for implementing nonroutine problems into their curriculum, this book defines and establishes practical strategies to develop students' problem-solving skills. The text focuses on the process skills necessary to solve nonroutine problems in mathematics and other subjects, with the goal of making students better problem-solvers both in and outside of the classroom.

Chapters present and define a curriculum of over 60 nonroutine problems in mathematics and other content areas, and explore the pedagogy to implement this type of curriculum consistent with the NCTM Standards and Principles to Action. Four different models of implementation are discussed, alongside a structured approach through seven difficulty levels (with examples), to ensure that every student, independent of their mastery of mathematics content, can improve their ability to solve nonroutine problems. It emphasizes to students how to transfer their problem-solving skills to other real-world areas, including increasing ecological awareness, appreciating diversity and addressing significant and meaningful problems in their life, school and community. The curriculum introduced in this book can be included as a component of a traditional four-year academic high school curriculum aligned with the Common Core Mathematical Practices, or as part of a one-year isolated required or elective mathematics course.

Based on extensive field-testing this approach has been effective in both traditional mathematics courses and math electives such as a course in Problem-Solving. This book provides the necessary guidance to allow each mathematics teacher to effectively integrate the approach in their classrooms. This book is ideal for secondary mathematics teachers of all levels, as well as teachers of mathematics electives.

Robert London is a professor emeritus at California State University, San Bernardino, USA, and was program coordinator for the MA in Holistic and Integrative Education and a mathematics educator. He also has 20 years of experience as a secondary mathematics chair and teacher, school director and elementary and middle school teacher.

Introducing Nonroutine Math Problems to Secondary Learners

60+ Engaging Examples and Strategies to Improve Higher-Order Problem-Solving Skills

Robert London

Routledge
Taylor & Francis Group

NEW YORK AND LONDON

Designed cover image: © Getty Images

First published 2024
by Routledge
605 Third Avenue, New York, NY 10158

and by Routledge
4 Park Square, Milton Park, Abingdon, Oxon, OX14 4RN

Routledge is an imprint of the Taylor & Francis Group, an informa business

© 2024 Robert London

ISBN: 978-1-032-48378-8 (hbk)
ISBN: 978-1-032-48377-1 (pbk)
ISBN: 978-1-003-39328-3 (ebk)

DOI: 10.4324/9781003393283

Typeset in Palatino
by codeMantra

Contents

Acknowledgments

Many people have contributed to the content of this book, either directly or indirectly. Here I will mention just four very direct contributors. First, my connection with the Project to Improve Mastery of Mathematics and Science (PIMMS) project, primarily through several summer fellowships between 1984 and 1995, was the primary impetus for this book, especially the 1984 summer that introduced me to a number of ideas for what I initially considered to be just enrichment problems for my Calculus class (see Chapter 1). I especially want to thank Dr. Robert Rosenbaum, formerly a mathematics professor at Wesleyan University, and Steven Leinwand, then State Mathematics Education consultant in Connecticut, for their leadership in that program, both in terms of financial support primarily from industry and the government and in terms of their clear support for our growth as mathematics educators. The initial impetus for my work with nonroutine problems was that first summer and the many thoughtful discussions we had. Second, I need to thank the Alden B. Dow creativity center's ten-week summer fellowship in Michigan and the support of the staff at their center, especially the director, Carol Coppage, that allowed me to expand my curriculum from eight problems just appropriate for high-ability mathematics students (e.g., calculus students) to a curriculum of 60 problems in seven strands for a four-year secondary mathematics curriculum. Third, I need to thank my students at Old Saybrook High School in Connecticut who field-tested the curriculum, especially my Problem Solving classes which allowed a whole year of instruction in a variety of problems. Fourth, as a professor at California State University, San Bernardino, through a small grant I was able to teach MATH 302, Problem Solving, to undergraduate liberal studies students fulfilling their mathematics requirement. The students in that course allowed me to further field-test the program, especially their ability to transfer the skills to both mathematics problems and meaningful real-life problems.

In terms of indirect support, I must first express my gratitude to my wife, Janet, who has supported me for more than 50 years, certainly a nonroutine problem well addressed! Also, I must thank my parents who allowed me to become the person I am rather than someone they wished me to be. I cherish their parenting which provided the support, environment and structure I needed to grow into a healthy and meaningful life. Finally, I must thank my daughter, Jenn, son-in-law, Marc and granddaughter, Audrey who have given me much joy in my life. Jenn, I reaffirm my promise to give you the role of parabola when this book is made into a movie.

Foreword

Problems – what a delight! Problems that engage our minds. Problems that stimulate our thinking and our emotions. Problems that teach and frustrate, that pique our curiosity and leave us feeling inadequate. All part of the universal problem-solving process – doing what we do when we really aren't sure what to do.

As technology – Google, Alexa, Desmos, AI, etc. – becomes more and more ubiquitous, an empowering secondary, pre-collegiate education must focus more and more on problem solving, modeling, reasoning and justification – the increasingly essential skills in today's world. Why are we still training the masses to rationalize denominators or practice synthetic division or prove trig identities or memorize (as opposed to select) geometric theorems or factor trinomial expressions with leading coefficients greater than one for weeks and weeks? Instead, why aren't we recognizing that replacing the obsolete chaff of mathematics with the nutritious wheat of nonroutine problems is more important than ever to ensure a literate citizenry and prepared workforce?

Bob London has fretted about, studied, played with, taught and written about problems and problem solving for his whole career. In this welcome volume, Bob presents, discusses guidelines for and provides mathematical and pedagogical supports for more than 60 diverse exemplar nonroutine problems that he has used successfully with high school and college students for years.

The clarion call for meaningful problem solving has been a centerpiece of mathematics curriculum and instruction recommendations for more than forty years. In *An Agenda for Action*, back in 1980, the National Council of Teachers Mathematics (NCTM) presented "problem solving must be the focus on school mathematics in the 1980s" as recommendation #1. When NCTM essentially launched the modern day educational curriculum standards movement in 1989 with *Curriculum and Evaluation Standards for School Mathematics*, "mathematics as problem solving" was Standard 1 in grades K-4, 5–8 and 9–12. For grades 9–12, for example, teachers and curriculum developers were advised that:

The mathematics curriculum should include the refinement and extension of methods of mathematical problem solving so that all students can:

◆ use, with increasing confidence, problem-solving approaches to investigate and understand mathematical content;

- ◆ apply integrated mathematical problem-solving strategies to solve problems from within and outside mathematics;
- ◆ recognize and formulate problems from situations within and outside mathematics;
- ◆ apply the process of mathematical modeling to real-world problem situations.

In 2010, despite its unfortunate politicization, first among the eight Standards for Mathematical Practice in the *Common Core Standards for School Mathematics* is "make sense of problems and persevere in solving them" elaborated with:

Mathematically proficient students start by explaining to themselves the meaning of a problem and looking for entry points to its solution. They analyze givens, constraints, relationships and goals. They make conjectures about the form and meaning of the solution and plan a solution pathway rather than simply jumping into a solution attempt. They consider analogous problems, and try special cases and simpler forms of the original problem in order to gain insight into its solution. They monitor and evaluate their progress and change course if necessary. Older students might, depending on the context of the problem, transform algebraic expressions or change the viewing window on their graphing calculator to get the information they need. Mathematically proficient students can explain correspondences between equations, verbal descriptions, tables and graphs or draw diagrams of important features and relationships, graph data and search for regularity or trends. Younger students might rely on using concrete objects or pictures to help conceptualize and solve a problem. Mathematically proficient students check their answers to problems using a different method, and they continually ask themselves, "Does this make sense?" They can understand the approaches of others to solving complex problems and identify correspondences between different approaches.

And most recently, in 2014, in *Principles to Actions: Ensuring Mathematical Success for All*, NCTM advocates for "implementing tasks that promote reasoning and problem solving" as one of the eight Mathematics Teaching Standards, arguing that "effective teaching of mathematics engages students in solving and discussing tasks that promote mathematical reasoning and problem solving and allow multiple entry points and varied solution strategies."

Unfortunately, while the aspirations are clear and widely advocated, too often classroom reality remains far from what is needed. Inadequate guidance, inadequate instructional materials, traditional assessments and the power of tradition all conspire to limit these changes.

In this book, however, we have a place to start. Consider the need to think outside of the box and to truly problem-solve with the following nonroutine, yet accessible task:

A census taker comes to the house of a mathematician and asks how many children he has and what are their ages. The mathematician replies that he has three children and the product of their ages is 72. The census taker replies that he has not been given enough information to determine their ages. The mathematician adds that the sum of their ages is the same as his house number. The census taker leaves but returns in ten minutes and tells the mathematician that he still does not have enough information to solve the problem. The mathematician thinks for a short while and then adds that the oldest child likes chocolate ice cream. The census taker replies that he has enough information and leaves. Question: What are the ages of the three children?

Bob points out that this problem, unlike the other problems in the book, is not meaningful; however, the solution to the problem contrasts the approach of a mathematician and the average student to attacking the problem, concretely illustrating the three steps in practice. I'll leave it to the reader to wrestle with this one for a while or continue on with the book to find an elegant solution. But the message of thinking, reasoning, changing perspective, collaborating with others and crafting a clear justification for a solution are all evident and in stark contrast to the all-to-typical memorization and regurgitation of mathematical procedures that essentially no one with a smart phone does anymore.

One will also find a broad range of types and content within this compendium and guide to nonroutine problems. You will play with, and learn to use, algebraic and geometric problems as well as those that are statistical, personal, practical and cultural. All requiring diverse and increasingly essential skills.

But it is the process of problem solving and the training of the mind in the use of these processes that this book offers that I find most helpful. Bob draws on his life's work and provides important details and insights into his belief that the essence of a nonroutine problem is that it is a problem at an appropriate level of difficulty that requires three steps to solve satisfactorily: problem orientation and recognition, trying something and persistence.

Perhaps the best way to transition from a guest's foreword to the authors work is to simply whet your appetite by quoting the passage that grabbed me early in this book and that I hope grabs and motivates you as well:

My intention in the body of this book is to give you a well-organized explanation of what a curriculum of nonroutine problems is and how to implement such a curriculum in your professional context....

Outcomes of implementing the curriculum: For example, instead of stopping when an obstacle is encountered, the student will persist. Instead of ignoring obvious contradictions or inaccuracies, the student will actively examine them. Instead of being intimidated by ambiguity, the student will tolerate the ambiguity. Instead of being satisfied with the first solution to a problem, the student will work on a problem until a more satisfactory solution is reached. Instead of staring at a problem that seems unsolvable or confusing, the student will try something until the problem naturally becomes clearer.

Go play. Go solve. Go learn. Go try with your students. Go forth and make a difference in the lives of the next generation.

Steve Leinwand, American Institutes for Research; Coauthor, NCTM's
Principles into Actions: Ensuring Mathematical Success for All

Foreword

Bob London has been a leading figure in the holistic education movement since the mid-1990s. With Sam Crowell at California State University in San Bernardino, he developed a master's program in holistic education; he was head of the Spirituality and Education SIG at AERA and chaired a major conference in holistic education in 2007. Now he has given us the first comprehensive holistic approach to mathematics education, perhaps the only example of a holistic, process-oriented comprehensive approach to mathematics education.

The concept of nonroutine problems is central to the book which Bob defines as "a problem at an appropriate level of difficulty that requires three steps to solve satisfactorily: problem orientation and recognition, trying something and persistence." In addition to mathematical problems, there are "nonmathematical" problems such as those related to the environment, increasing appreciation of diversity and meaningful problems in the student's life, school and community. Much of the material for this book comes from a course Bob taught on Problem Solving at Old Saybrook High School in Connecticut for twelve years. An example of a non-mathematical problem that he gave to students was to plan a trip to New York (about 100 miles from the school) that needed to be educational, enjoyable and inexpensive. Students had to employ the three steps to solve the problem. In the second half of the course, students had to identify and start solving two significant problems in their life. For example, these included problems such as a major purchase such as a used car, obtaining employment, budgeting or selecting a post-secondary institution.

Nonroutine problems are open-ended, require the student to "use one or more mathematical problem solving strategies such as finding a pattern and generalizing, generating and organizing data, manipulating symbols and numbers or reducing a problem to an easier equivalent problem." They cannot be solved in a few minutes and require persistence to complete. It is the last feature of nonroutine problems that makes it particularly relevant to today's world. With Google, ChatGPT and other tools on the internet students can quickly find answers to problems. However, life often presents complex problems that cannot be solved easily or found on Google and require persistence. Bob's model helps not only in problem solving but also in developing an important life skill that is relevant to the development of character. Bob argues that working with nonroutine problems develops mathematical maturity which includes:

> For example, instead of stopping when an obstacle is encountered, the student will persist. Instead of ignoring obvious contradictions or inaccuracies, the student will actively examine them. Instead of being intimidated by ambiguity, the student will tolerate the ambiguity. Instead of being satisfied with the first solution to a problem, the student will work on a problem until a more satisfactory solution is reached. Instead of staring at a problem that seems unsolvable or confusing, the student will try something until the problem naturally becomes clearer.

I would argue that dealing with "ambiguity" is not only a sign of mathematical maturity but personal maturity as well.

This book presents different ways the material can be included in the curriculum from a 3- to 4-year curriculum for a secondary school, to a one-year course on problem solving, to an "integral component of an interdisciplinary curriculum." The book is filled with practical suggestion of how to implement ideas presented and includes more than 60 nonroutine problems that teachers can use in their classes. Seven levels of support and assistance are presented "to provide a general instructional sequence to help students move effectively to the objective of being able to solve nonroutine problems without significant scaffolding." At the first level, the teacher "defines a nonroutine problem and describes a good solution(s) to the problem. Students are required to summarize the problem and solution." At the last level, students generate and define nonroutine problems important to them individually and solve each problem.

This book is an important addition to the literature on holistic education as there has been very little literature on holistic approaches to mathematics. This book is a gift to teachers seeking materials and guidance on how mathematics can be made relevant to their students and the problems they face in their lives. Bob's comprehensive approach to problem solving also helps students grow and become thoughtful adults.

John (Jack) Miller
Professor, University of Toronto

1

Introduction

What is essential in mathematics education? This particular question has driven my research and teaching for over 40 years. In my opinion, any curriculum that claims to address this question must provide the student with the tools to effectively address significant problems in that student's life. This book concerns an approach to teaching mathematics that facilitates the student seeing the connection between the process of solving nonroutine mathematics problems and solving significant problems in the student's life. The approach is embedded in a curriculum or sequence of nonroutine problems. Curriculum in the sense of a process with guidelines that can be integrated into a number of formats, including an independent course; a sequence of traditional secondary mathematics courses; an undergraduate course and an innovative, interdisciplinary approach.

My intention in the body of this book is to give you a well-organized explanation of what a curriculum of nonroutine problems is and how to implement such a curriculum in your professional context. While I believe this approach is essential to allow you to implement the ideas suggested in the book, it omits the story and the why of this book. The body of this book well documents the *implications* for instruction of over 25 years of work and research; however, it does not give you a sense of the discoveries, the failures, the successes, the realizations of the process that have convinced me of the significance of a curriculum of nonroutine problems. In this introductory chapter, I want to convey a sense of why I wrote this book and why I

DOI: 10.4324/9781003393283-1

believe the approach presented is significant, hopefully thereby deepening your connection with the essence of the approach.

Basically, the story starts in the 1984–85 school year. I had been mathematics department chair and teacher at Old Saybrook High School in Connecticut since 1981. Before moving to Connecticut, I had spent ten years in Massachusetts, earning my doctorate from 1971 to 1975 and working in a variety of positions. In the years from 1971 to 1984 the focus of my work in curriculum development was problem solving with a focus on improving students' ability to use mathematical problem solving strategies (e.g., problem reduction, seeing a pattern, guess and check) in a variety of mathematical content contexts. A PIMMS (Project to Increase Mastery of Mathematics and Science) Summer Fellowship in the summer prior to the 1984–85 school year initiated a change in my focus. That summer I was introduced to a number of interesting mathematics problems. I decided to use some of the problems and others I created or adopted as enrichment problems in my calculus class. At the beginning of the school year, I did not see the problems as a "curriculum" or even as a significant addition to the course. I thought they were challenging problems that the students might enjoy, while learning some interesting mathematics. Feeling the pressures of coverage of the typical calculus course, I allowed little class time for the problems. I took a small portion of a class to introduce a problem, let them work on the problem outside of class for approximately two weeks, collected their solutions, and, after reading and assessing their work, spent a portion of a class processing their work and discussing various solutions and strategies and, if appropriate, discussing the significance of the problem. A few of the problems were excellent introductions to concepts in the course and therefore were further discussed.

So far, nothing worth writing a book about! It is fair to say that I have had my share of successful interactions with students, effective classes and units and rewarding courses. However, in a long career of teaching, there are a very few classes that have a special significance and that I can still remember clearly. One of those classes occurred that year. The students had finished their work on the eighth nonroutine problem and handed in their solutions. When I finished reading their papers, I suddenly realized that they had made a quantum leap in their understanding of the process of problem solving since the beginning of the school year. I have described this leap as follows:

> The significance of the curriculum can probably be best described by reporting the observed effect on the mathematical maturity of the students. It is as if the student has been transformed mathematically! Instead of acting in all the ways that we normally attribute to most high school students, the student acts similarly to a 'mathematically

mature' person. For example, instead of stopping when an obstacle is encountered, the student will persist. Instead of ignoring obvious contradictions or inaccuracies, the student will actively examine them. Instead of being intimidated by ambiguity, the student will tolerate the ambiguity. Instead of being satisfied with the first solution to a problem, the student will work on a problem until a more satisfactory solution is reached. Instead of staring at a problem that seems unsolvable or confusing, the student will try something until the problem naturally becomes clearer.

(London, 1989, p. 1)

At the same time, I "saw" what each student needed to complete or solidify the leap and "saw" individual problems to assign each student to help them complete the process. I can still picture most of the students in the class and the assignments I gave them.

What I want to emphasize here is that it was this unexpected result that convinced me that I had stumbled onto something significant, not my own previous work or the work of others (of course, those factors facilitated the event happening!). That experience encouraged me to reflect on what it was I did that had been so successful. The result of the reflection was the development of the definition of a nonroutine problem and a curriculum of nonroutine problems, and the book *Nonroutine Problems: Doing Mathematics* (1989). That short book refined my work with the 1984–85 class and presented a curriculum of ten nonroutine mathematics problems appropriate for students with at least an Algebra 2 mathematics background. That marks the end of the first phase of the story.

Before moving to the second phase, I need to give you a sense of what is meant by a nonroutine problem and illustrate the need for a curriculum of nonroutine problems. I hope this digression will give you a framework to better appreciate the rest of the story in this introduction. Of course, these topics are treated in much more detail later in this book. For here, the essence of a nonroutine problem is that it is a problem at an appropriate level of difficulty that requires three steps to solve satisfactorily: problem orientation and recognition, trying something and persistence. Here, I will cite one example from my early work with calculus students to illustrate that even our best students generally lack an appreciation of the significance of these three steps: The first nonroutine problem I give my calculus students is calculating the area under the curve $y = x^2 + 2$ between $y = 0$, $x = 0$ and $x = 3$ (see Figure 1.1).

Despite directions to determine the area as best they can, the great majority of the students calculate the area as 15 1/2 square units by "replacing" the curve with the straight lines connecting (0,2) and (1,3), (1,3) and (2,6) and

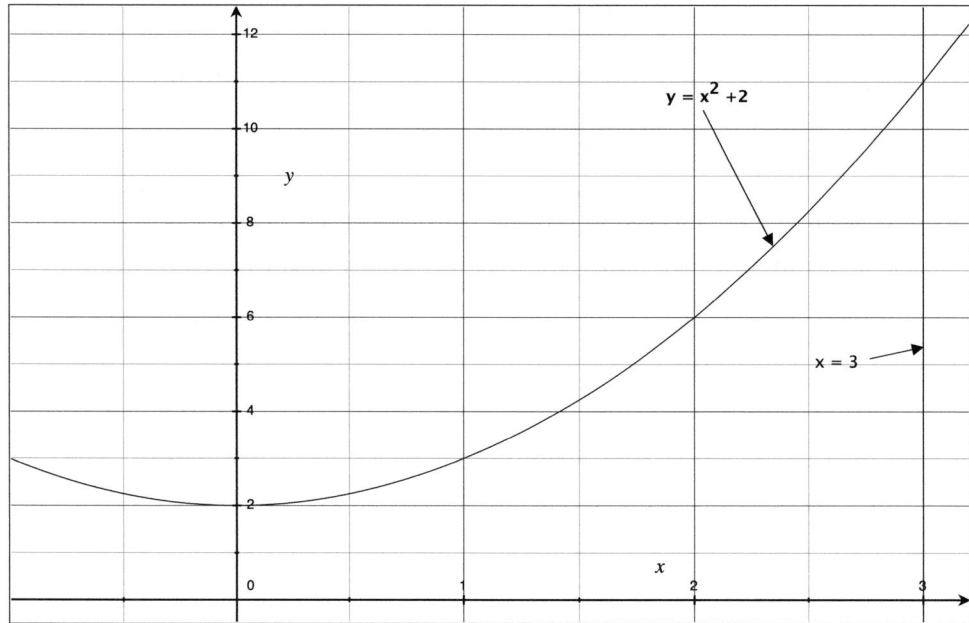

Figure 1.1 The area under the curve.

(2,6) and (3,11) and then dividing the area into triangles and rectangles (or trapezoids). The actual area is 15 square units. These students fail to recognize that their calculations could be more accurate by simply increasing the number of divisions (e.g., a change from three divisions to six divisions gives a result of 15.125 square units). From having worked with these students, I believe they simply do not have experience solving problems that require them to persist and evaluate the adequacy of their solutions. Certainly typical textbook problems do not require that skill. In terms of the first step, the students fail to make an initial connection with the solution; i.e., they have no clear sense at the beginning of the problem what an adequate solution (the best approximation of the area) looks like. Notice their solution is inadequate even though the strategy they picked (basically the trapezoid method) is excellent and, in fact, the basis of the tool of integration in calculus. I believe most of the students, typically high achievers, would have been willing to spend the time necessary to persist but simply did not have experience solving problems that require them to recognize that a problem does not always have a straightforward solution, despite my directions indicating the need. In contrast, I believe that if I gave these students this problem at the end of the curriculum, the majority would have persisted to a more adequate solution.

Back to the story. The second phase concerns the question that interested me once I felt satisfied with the curriculum for above average students: Can all students learn how to solve nonroutine problems? Fortunately, I had an excellent laboratory to begin to explore this question. Each year at Old Saybrook High School I taught a course Problem Solving, a full year course that I initiated that primarily served seniors who had no other viable alternative (in their opinion) for a final required mathematics course. This course always included a range of students, from students who had poor mathematics ability and had barely got by in our lowest level courses, to above average ability students who were not achieving at a level consistent with their potential; from students who were motivated to learn and willing to work hard, to students who at least outwardly seemed to have no interest in school work; and from students who had no or few positive experiences in their history as a mathematics student, to students who enjoyed mathematics. In previous years, my emphasis in the course had been on improving the student's ability to use the major problem-solving strategies. I decided to spend a significant portion of that course on improving the students' ability to solve nonroutine problems.

This was the only course except AP Calculus that I taught consistently each year. I enjoyed the class immensely. I will always remember the typical first class for this course. The students would walk in and be surprised to see in the same class so many students so well known for their lack of interest in succeeding academically in school. I remember the looks or remarks of students indicating the unstated question to me: "Did I know what I was getting myself into?", some asking the question as a challenge, others perhaps out of compassion. I did know into what I was getting myself, at least after the first year or so. I must say I welcomed these students because, in my opinion, for the most part, they had a better sense than the average student of the craziness of what we were teaching them in the name of mathematics education. Perhaps they were not successful creating something more meaningful in their life, but at least they were not participating in something for which they could see no rationale or meaning. I liked a class that would let me know clearly if in their opinion I was on track or not. Of course, the class was much more complex than I am indicating. Many students felt they should be doing better even though part of them would not allow them; some of them had self-destructive tendencies that interfered even when the curriculum was meaningful; others bought into my vision of the course and worked hard the whole year. What is relevant is that I felt it was my responsibility to teach them something that was convincing to them as being meaningful, and I knew their reactions to my teaching would force me to keep my attention on that responsibility.

Perhaps it was the pressure to create a meaningful curriculum for these students that forced me to consider what was meaningful about a curriculum of nonroutine problems. This search for meaning was not always conscious, but rather usually the result of what was happening or not happening with my Problem Solving class. In any case, after a few years of experimentation I had developed an approach that I began to feel was resulting in an effective, meaningful curriculum. The major realization on my part was that the three steps of solving a nonroutine problem involving mathematical content were the same steps that were effective in solving significant problems in one's day-to-day life and in one's occupation. I saw how I used the steps to solve significant problems in my life and came across numerous examples of other people in a variety of occupations and situations using the steps to solve significant, meaningful problems. I saw that the ability to use the three steps in one's life could significantly improve the quality of one's life. This realization affected the curriculum for my Problem Solving class in a number of ways. For example, I began to include an introductory unit which defined the three steps, gave numerous examples of people in a variety of "nonmathematical" situations using the three steps to solve significant problems in their life and required students in groups to research additional examples. I also included a variety of problems in "nonmathematical" contexts that not only required the three steps of a nonroutine problem to solve, but also were seen by the students as being meaningful. For example, each year the class planned and took a trip to New York City (about 100 miles from our school) that was enjoyable, educationally valuable and inexpensive – a difficult task! Not only did the students enjoy the trips, but also they could see how they could use the three steps in their lives to plan trips (sometimes in the curriculum as a follow-up problem they were required to plan an actual individual or small group trip). For example, for most students the New York trip was significantly less expensive than other school trips and much more enjoyable.

A significant improvement in the curriculum was a requirement in the second half of the course that they had to identify and solve (or begin the process of solving) at least two significant problems in their life. Typically, students selected problems involving a major purchase (e.g., used automobile, stereo system or bicycle), obtaining employment (e.g., a well-paying, interesting summer job), budgeting (e.g., budgeting to buy an automobile) or selecting a post-secondary institution. Two examples that indicate the potential of this approach are taken from two vocational students of poor mathematics ability who took my Problem Solving course. The first student had a lawn care business with four clients with large lawns. He had difficulty organizing the business and was ready to give it up. His problem was to effectively organize the business. To gather information, he interviewed his clients and three lawn care professionals. His initial conclusions included (1) plan a schedule to complete the lawns in four days, allowing for bad weather and other complications and (2) schedule your largest lawn early in the week. Based on these conclusions and other data, he devised a tentative plan that he checked with his clients and a fourth lawn care professional. He successfully implemented the plan and was very satisfied with the results. The second student picked a very personal problem: he was concerned about his inability to relate well with other students, particularly his inability to make close friends. He investigated a number of options and decided to see our school psychologist. I saw her the next fall and she indicated that he continued seeing her after graduation through the summer, making significant progress. To me, these two examples indicate the power of a curriculum of nonroutine problems. These two students were able to define a significant problem in their life and apply the steps a mathematician uses to solve a difficult problem to effectively solve their problem. I claim that mastery of this type of problem solving is what is essential in mathematics education and most useful to the great majority of students.

The third phase of this story started when I was awarded a ten-week summer Alden B. Dow Creativity Fellowship in 1992 to write a draft of a four-year secondary curriculum of nonroutine problems for the average or below average ability mathematics student. I and three other fellows were given individual apartments on the campus of Northwood College in Midland, Michigan, and allowed to spend our entire time working on our projects, isolated from friends, family and other distractions of our normal day-to-day lives. Each of us was motivated to make good use of this rare opportunity to focus almost exclusively on creating, and we supported each other working on our projects. In addition, for me it was an opportunity to reflect on the curriculum without the pressure of planning an actual class each day.

The primary product of the summer was a curriculum consisting of 60 nonroutine problems, 15 a year involving several strands including strands that increased the student's appreciation of diversity, improved the student's ecological awareness and required the students to define and solve significant problems in their lives. The next three school years were spent field-testing the curriculum, revising many problems, creating new problems and working with other teachers to develop a curriculum we felt was appropriate to implement in our school. A two-summer PIMMS Fellowship with a focus on preparing leaders in technology in mathematics education in Connecticut allowed me to further integrate the use of technology into the curriculum.

The fourth phase of this story was marked by my accepting a position as an assistant professor in education at California State University, San Bernardino, starting in the fall of 1995. As much as I enjoyed teaching high school mathematics I was ready for a change. While the daily demands of high school teaching had forced me to "discover" the significance of nonroutine problems and face the challenges of implementing a curriculum of nonroutine problems as well as the required content objectives of the academic courses I taught, these same demands were no longer consistent with what I felt I needed. My 14 years at Old Saybrook High School had been extremely fruitful, resulting in much curriculum that I felt was effective. However, I had volumes of materials that I felt would be useful for other teachers, but I had not had the time to publish them in a useful form. In addition, I felt that I had been successful in developing most of the curriculum in which I was interested and that it was time for a shift to more reflection and writing, and less time in the classroom. Probably the Alden B. Dow Fellowship had been a turning point in my attitude – I realized that I spent most of my time that summer writing up creative ideas (over 500 pages!) that had occurred in previous years while teaching when I had no time to write up the ideas in a reflective manner. I wanted the time to reflect on the curriculum, revise it and work with other teachers to field-test it when appropriate. As a professor of education, I continued to develop the curriculum, field-tested the approach with some mathematics teachers and applied the approach in teaching an undergraduate mathematics course in problem solving.

Perhaps a key event was in the spring of 1997. While I had written a number of articles on the curriculum, I had not really begun serious work on this project. I was not clear for whom and in what form I wanted to publish the curriculum – as part of a textbook series, as supplemental material or what? At that time, Sam Crowell, a colleague, asked me to proofread the manuscript for his soon to be published book *The Re-enchantment of Learning* (Crowell et al., 1998, draft manuscript). I agreed. I was immediately struck by the introduction to the book. The introduction seemed to effectively communicate the essence of the work and what was hoped the reader would get out of the book. In part, he wrote:

> Quite simply, we want you to know that this is a book from the heart. It represents more than just our knowledge and experience; it is in some ways a story of our own deeply held commitments. It is also a story of what the teachers we have worked with over the years have taught us. It is a book that we want you to experience and not just absorb. Our hope is that it will become an integral part of your professional and personal life.
>
> (draft manuscript, no page number)

His words resonated with what I wanted to communicate to my readers, perhaps with the addition that this is a story of not only what my fellow educators have taught me, but also more importantly what my students have taught me. His introduction helped me focus on what I wanted to communicate in this book. The purpose and tentative outline of the book soon became clear. Simply put, I feel I have stumbled upon an approach to teaching mathematics that is significantly more meaningful than what we have been doing in the past, even those of us that have always emphasized problem solving. I believe that an understanding of, and the ability to apply, the three steps of a nonroutine problem to solve meaningful problems in one's life is the essence of what is most meaningful in the mathematics curriculum. In fact, elsewhere I have described a model for an alternative school consistent with the needs of the 21st century and argued that a significant component of the model should be a curriculum of nonroutine problems focusing on identifying and solving nonroutine problems in the local community (London, 1996).

This feeling of the significance of this approach demands that I outline clearly in this book my best guess at how a curriculum of nonroutine problems should be implemented. I will try to convey the essence of what I see as a holistic approach to the teaching of mathematics, realizing that I cannot do justice to all the specifics of the curriculum within the space limitations of this book, as well as the wide variety of professional contexts of the readers. In brief, the approach I suggest requires you to rethink what the mathematics curriculum should look like and to eliminate or deemphasize many of the things we have assumed for many years are important in the mathematics curriculum. In other words, I am going to suggest some radical changes in how we teach mathematics. At the same time, as a public school teacher for many years, I know that the suggested changes cannot be easily implemented in the majority of schools. Therefore, I believe it is equally important that I suggest how teachers in a variety of settings can implement a curriculum of nonroutine problems consistent with their school setting and their beliefs, especially individual public school teachers responsible for teaching the content of a traditional academic course (e.g., algebra 1, geometry). In summary, I have two goals in writing this book. First, to present unequivocally what I believe a portion of the mathematics curriculum should look like, hoping that at least a few courageous teachers in favorable settings can implement this vision or their version of this vision. Second, to give every mathematics teacher, in whatever setting, the information, advice and support to implement a curriculum of nonroutine problems in their classroom or school, yet consistent with the perceived limitations in their professional context.

I tried to organize the remainder of the book in an order that is comprehensive, providing guidance and suggestions for a wide variety of contexts. Specifically, Chapters 2 and 3 discuss basic information necessary to build a foundation for implementing a curriculum of nonroutine problems. Chapter 2 provides a conceptual framework for understanding the concept of a curriculum of nonroutine problems and outlines the basic pedagogical principles key to implementing such a curriculum. Chapter 3 describes in detail ten introductory nonroutine mathematics problems that provide the necessary scaffolding to facilitate students becoming comfortable solving more difficult nonroutine problems. Chapters 4–6 outline nonroutine problems related to the traditional mathematics curriculum, algebra, geometry, precalculus and calculus, prediction, probability, estimation and number theory. Chapters 7–9 concern additional nonroutine problems for which the content focus is not typically considered mathematical, including the following strands: introductory nonroutine problems for building community; developing ecological awareness; increasing appreciation of diversity; and individual and class-generated problems that are considered meaningful. Chapter 10 discusses methods to implement a curriculum or sequence of nonroutine problems in a variety of professional settings, including as a coordinated component of a traditional four-year secondary mathematics curriculum; in the context of one teacher for a one-year academic or elective secondary mathematics course; an undergraduate course for liberal arts majors; and as part of a curriculum integrated into the entire high school curriculum. Chapter 11 discusses additional problems and resources.

I have tried to be complete in the suggestions in the remainder of the book, but one essential suggestion needs to be emphasized – allow yourself to be open to the richness of the problems, to the variety of solutions and to the opportunities for growth that the students' efforts will create.

References

Crowell, S., Caine, G., and Caine, R. (1998). *The re-enchantment of learning*. Tucson, AZ: Zephyr Press.

London, R. (1989). *Nonroutine problems: Doing mathematics*. Providence, RI: Janson Publications.

London, R. (1996, March). *An alternative school model*. Paper presented at the Association for Supervision and Curriculum Development National Conference, New Orleans.

2

Conceptual framework and pedagogy

In this chapter, I will provide a framework for understanding the concept of a curriculum of nonroutine problems. I will discuss the definition of a nonroutine problem, including an explanation of a rationale for the definition, and then define a curriculum of nonroutine problems and illustrate the concept in four different professional contexts. I will also briefly discuss some of the implications for pedagogy in implementing a curriculum of nonroutine problems. I will note that the framework is based on my actual experience developing and implementing such a curriculum, and my ongoing processing, assessment and reflections on that experience, rather than an extensive review of the relevant literature. Of course, that process has been significantly affected by my professional work (e.g., London, 1976) and my continuous study since 1971 of others' work, especially in areas such as problem solving, constructivism, holistic education and process-oriented curriculum. I believe such an approach based primarily on my actual experience is more honest and more reader friendly, hopefully resulting in a clearer understanding of the approach. Finally, I will add that I will discuss in detail my interpretation of the relationship of this approach to the current emphasis on the Common Core Mathematical Practices (CCMP) – in short, I support the theoretical framework for the CCMP, and I will describe how the pedagogy described for the curriculum supports that approach.

DOI: 10.4324/9781003393283-2

Definition of a nonroutine problem

In recent years, educators have made a major effort to improve the mathematical problem solving abilities of students. We have seen the improvement of problem solving skills identified as an important focus of the National Council of Teachers of Mathematics (NCTM) and other professional organizations. In recent years, the professionally aware mathematics teacher has been flooded with ideas, techniques and materials to use in the classroom to improve problem solving skills, especially in terms of the CCMP.

This book is concerned with a type of problem not typically included in even the improved materials since the 1980s. I call this type of problem an open-ended nonroutine problem or, more simply put, a nonroutine problem, which, for the purposes of this book, has the following characteristics:

(1) The problem requires three steps to complete: problem recognition and orientation, trying something and persistence.
(2) The problem is open-ended; that is, it allows for various solutions.
(3) The problem requires the student to evaluate a variety of potential approaches to the problem and select one or more to pursue. Typically, a good solution requires the student to use one or more mathematical problem solving strategies such as finding a pattern and generalizing, generating and organizing data, manipulating symbols and numbers or reducing a problem to an easier equivalent problem.
(4) Every student is able to "solve the problem." Of course, the quality of different student solutions will vary, but students will be able to confront the problem and generate a solution consistent with their ability and efforts. Stated differently, the problem will be challenging, but not unnecessarily frustrating.
(5) Each problem cannot be solved straightforwardly in a few minutes, typically requiring at least a few hours work over at least a week's time, as well as requiring reflection on the process of solving the nonroutine problem.

If a specific student can solve a given nonroutine problem quickly then, by definition, the problem is not a nonroutine problem for that student.

In a sense, a nonroutine problem in mathematics corresponds to an essay requiring creative writing in language arts in that the problem requires a significant investment of time; most students can complete the assignment and the problem requires higher-order thinking skills. To push the analogy one step further, just as writing an essay can be considered the step beyond writing sentences and paragraphs, solving nonroutine problems can be considered the step beyond solving problems (or sets of problems) involving one or two strategies that can be solved in a relatively short period of time.

It is necessary to clarify the fourth characteristic of a nonroutine problem: Every student is able to "solve the problem" in the sense of trying something and perhaps persisting until there is a change in the level of understanding. In my opinion, given the realities of the level of understanding of mathematical content of the typical high school student, it is inappropriate initially to include nonroutine problems that require significant mathematics content prerequisite skills, *if your objective is to improve the student's ability to solve nonroutine problems*. In early work (London, 1976), I developed a six-fold classification system for what I labeled as process-oriented curriculum. For example, the classification system discriminated between curriculum that primarily focuses on process skills (e.g., the three steps of solving a nonroutine problem) and curriculum that primarily focuses on subprocesses (e.g., specific problem solving strategies or heuristics such as finding a pattern, guess-and-check, problem reduction) or content (e.g., algebra or geometry). Using the classification system, the curriculum of nonroutine problems is a combination of process-oriented curriculum that primarily focuses on process skills and process-oriented curriculum that primarily focuses on process skills but takes into account content. To clarify the significance of this classification, consider the following two problems from Schoenfeld (1985, pp. 15–6) that are examples of problems requiring the three steps of nonroutine problems, but not satisfying the criteria that every student is able to "solve the problem":

(1) Given two intersecting lines and a point P marked on one of them, show how to construct, using straightedge and compass, a circle that is tangent to both lines and that has the point P as its point of tangency to one of the lines.
(2) Given triangle ABC, show that it is always possible to construct, with straightedge and compass, a straight line that is parallel to line segment AB and that divides triangle ABC into two parts of equal area.

It is not implied that these two problems are inappropriate for high school students, but that, because of the content prerequisite skills assumed, these problems would not provide a good context for focusing on the skills involved in solving nonroutine problems for most students. In contrast, the curriculum of nonroutine problems was constructed so as to minimize the mathematical content prerequisite skills required of the student, especially in the initial stages of the curriculum. Therefore, using Schoenfeld's terminology (1985), this approach does not directly attempt to improve the category he labels resources ("in short, the mathematical knowledge… that the individual is capable of bringing to bear on a particular problem" (p. 17)). The approach in this book for a traditional public school accounts for less than 20% of the secondary mathematics curriculum, and a number of the problems directly or indirectly introduce significant content in the curriculum. In addition, field-testing has supported the hypothesis that a curriculum of nonroutine problems can be effectively integrated into the mathematics curriculum. In fact, it is hypothesized that by the third or fourth year of the curriculum, students would be better prepared to tackle problems such as Schoenfeld's as a natural part of the mathematics curriculum.

The definition of the three steps of problem solving is consistent with other definitions of the steps of mathematical problem solving (e.g., Polya, 1962) and a properly structured curriculum of nonroutine problems is consistent with constructivist theory (e.g., Brooks, 1993, Ward, 2001, Hickey, 2001). Previous work (e.g., London, 1989, 1993, 2004, 2007) supports the hypothesis that:

(1) The ability to solve nonroutine problems is an essential skill in mathematics education in the sense that these skills can be applied by students to solve significant problems in their life.
(2) Even most above-average ability students in mathematics need significant instruction in this objective.
(3) Most students need an integrative approach emphasizing problems from a variety of fields, including problems meaningful to them, to insure transfer to significant problems in their personal life.
(4) Successful instruction is feasible within a curriculum consistent with the NCTM standards. In addition, a curriculum of nonroutine problems can be easily integrated into other contemporary approaches to curriculum such as project-based learning (Curtis, 2002), creating communities of mathematical inquiry (Peressini, 2000), rich tasks (Moulds, 2004), place-based education (Sobel, 2005) or WebQuests (March, 2004).

Rationale for a curriculum of nonroutine problems

Why are open-ended nonroutine problems appropriate for inclusion in the secondary mathematics curriculum? Four reasons will be discussed in this section:

(1) Each nonroutine problem gives the students practice with three important steps of higher-order problem solving which are

typically not covered or emphasized in the traditional mathematics curriculum and are essential to doing mathematics: problem recognition and orientation, approaching a difficult or ambiguous problem by trying something or generating data and persisting until reaching a satisfactory solution.

(2) A sequence of nonroutine problems gives students experience with additional problem solving skills such as finding a pattern and generalizing, developing algorithms or procedures and describing them, generating and organizing data, manipulating symbols and numbers, and reducing a problem to an easier equivalent problem.

(3) Most students who have successfully completed a sequence of at least ten nonroutine problems demonstrate a mathematical maturity rarely observed in high school students.

(4) A curriculum of nonroutine problems supports students transferring the skills to solve meaningful problems in their life.

Three steps of problem solving

As stated above, one reason to include nonroutine problems in the curriculum is that each problem gives the students practice with three important steps of higher-order problem solving: problem recognition and orientation, approaching a difficult or ambiguous problem by trying something or generating data and persisting until reaching a satisfactory solution. In this section, I will attempt to clarify the three steps first through an exploration of the Census Taker Problem, a problem I explore with students when introducing a curriculum of nonroutine problems. In addition, I will cite some examples from students' work on nonroutine problems and discuss some other relevant examples from my experience as a mathematics teacher.

Most of the examples I use in the introductory unit of the curriculum to clarify the steps of a nonroutine problems are based on a variety of real-life examples of people solving significant problems in their life or profession using the three steps. In contrast, the Census Taker Problem is a recreational problem with no readily discernible connection with a real-life problem; however, I have found that the contrast between how mathematicians and typical secondary mathematics students approach the problem very helpful in explaining the three steps to the average student. I was introduced to the Census Taker Problem at a presentation by Krulik and Rudnick, authors of

Problem Solving: A Handbook for Teachers (1980), at a regional NCTM convention. They gave us the following problem to solve:

> A census taker comes to the house of a mathematician and asks how many children he has and what are their ages. The mathematician replies that he has three children and the product of their ages is 72. The census taker replies that he has not been given enough information to determine their ages. The mathematician adds that the sum of their ages is the same as his house number. The census taker leaves but returns in ten minutes and tells the mathematician that he still does not have enough information to solve the problem. The mathematician thinks for a short while and then adds that the oldest child likes chocolate ice cream. The census taker replies that he has enough information and leaves.

The problem is to determine the ages of the three children. When I read this problem to students, I ask them if they can identify two components of the problem that seem to be obstacles to finding a solution. Students are able to identify (as a class) two obvious obstacles to solving this problem: we do not know the house number and what does the fact that the oldest child likes chocolate ice cream have to do with determining the ages? At this point, I draw students' attention to the fact that even a mathematician would see the same two obstacles; i.e., at this point in the process the typical mathematics students are "equal" to the mathematician; a mathematician would not know some magical formula to solve this problem and would be similarly puzzled! However, there is a difference in how they might feel at this point – for example, average students might be quickly convinced that the problem is beyond their skills or quickly answer the problem with a "solution" clearly invalid. In contrast, a person comfortable with the strategy of approaching an ambiguous problem by trying something or generating data will generally start this problem by trying something, *despite the ambiguity of the problem.* For example, one could start by generating all the possible products of three whole numbers that equal 72, even though it may be totally unclear if or how this would help:

$1 \times 1 \times 72$
$1 \times 2 \times 36$
$1 \times 3 \times 24$
$1 \times 4 \times 18$
$1 \times 6 \times 12$
$1 \times 8 \times 9$

$2 \times 2 \times 18$
$2 \times 3 \times 12$
$2 \times 4 \times 9$
$2 \times 6 \times 6$
$3 \times 3 \times 8$
$3 \times 4 \times 6$

Indeed, starting this way combined with the third step of problem solving (persistence) can lead to the solution. At this point, I will not reveal the remainder of the solution. The point here is the same that Krulik and Rudnick made; that is, they drew our attention to the basic difference between the ways mathematics teachers and students approach this problem. They pointed out that even though only a fraction of the mathematics teachers correctly solved the problem in the given time, the great majority of the teachers had one characteristic in common – they all tried something, using pencil and paper. In contrast, they reported that students when given this problem typically do not write or try anything significant. Subsequently I have used this problem with several classes to introduce the three steps of problem solving and can verify that students typically do not fully try something.

The third step in problem solving emphasized in nonroutine problems is persisting until reaching a satisfactory solution. In the census taker problem, trying something by itself does not necessarily guarantee a solution. An example of persistence in this problem is the students who do not give up when they realize that the listing of all the possible products of three numbers equal to 72 does not lead directly to the solution and does not even seem to help clear up the ambiguity. Rather, a person who persists might eventually take the key step of listing the sums of each of the sets of three numbers whose product is 72, all the possible house numbers:

$1 \times 1 \times 72 = 74$
$1 \times 2 \times 36 = 39$
$1 \times 3 \times 24 = 28$
$1 \times 4 \times 18 = 23$
$1 \times 6 \times 12 = 19$
$1 \times 8 \times 9 = 18$
$2 \times 2 \times 18 = 22$
$2 \times 3 \times 12 = 17$
$2 \times 4 \times 9 = 15$
$2 \times 6 \times 6 = 14$
$3 \times 3 \times 8 = 14$
$3 \times 4 \times 6 = 13$

If one looks at this list, it is fairly likely that one would notice that there are only two sums that are identical: $2 \times 6 \times 6 = 3 \times 3 \times 8 = 14$. Now the original problem becomes easier; first, the house number must be 14; otherwise, the census taker would have been able to finish the problem (e.g., if the house number is 15, then the children must be 2, 4 and 9). Then the fact that there is an oldest child indicates that the two oldest children cannot be twins; therefore, the children must be 3, 3 and 8. Students often point out that the twins are not born exactly the same time. This observation is an excellent opportunity to point out that the context indicates that the census taker did not believe that was relevant, given the obvious implication of the clue from the mathematician.

To further clarify the three steps, I will discuss some additional examples from my students' work on nonroutine problems. The first step of problem recognition and orientation consists of three components: (1) a sense or attitude that the problem is solvable with the right type of effort (or at least the problem is worthy of attempting to solve); (2) a connection with step two; i.e., consideration of a variety of strategies and a choice of a strategy to pursue and (3) a connection with step three; i.e., a sense of what a good solution might look like and some sense of the obstacles to a solution. I will note that in practice this connection to step three is more a matter of realizing that a specific solution is not acceptable and that there is a need to persist. Problem recognition implies that the problem solver realizes that a particular solution is not satisfactory. An example that made a strong impression on me was in a calculus class I taught. To introduce a new technique, I asked the students to attempt an integration problem necessarily involving a new technique. To my surprise, most students solved the problem, of course, incorrectly! There was no awareness that this was a problem they could not solve with their present techniques!

A second example of the lack of this step involves quantitative comparisons. If they have not had instruction concerning this type of problem, many students will answer certain questions quickly and believe that they answered the question correctly. For example, if they are given that $a > b$, they will indicate that $a^2 > b^2$, not realizing that their solution is inadequate and that they should consider cases such as $a = 3$ and $b = -5$.

Similarly, a student approached determining the area of an irregular closed curve drawn on graph paper by calculating the average width times the length and ending up with an answer of 286.2 square units. Then the student calculated the area similarly, but by determining the area of the outside first and then subtracting from the total area. Using this method, the student obtained an area of 313.6 square units. The student's final solution was the average of the two values. In both calculations, the student generated and organized a good amount of data. However, the student obviously did not evaluate the adequacy of the answers. If the techniques were valid and carried out well, how could the student calculate two answers so different? What was needed was persistence to discover the minor mistakes that threw the calculations off. The reader might be thinking that perhaps the student was unwilling to take the additional time to do the needed calculations. My experience indicates that in general the students are not avoiding work, it just does not occur to most of the students that they could or should persist in the problem. After all, have any of the typical textbook problems they have answered over the years required persistence? I believe that very few if any have encouraged or required that type of thinking.

I want to give one more example of the third step by revisiting an example from Chapter 1 that hopefully not only clarifies the step, but also illustrates that even the best secondary mathematics students have difficulty (without instruction) with this step. The step of persistence implies having a connection with what a satisfactory solution looks like. For example, persistence implies that the problem solver realizes that a particular solution is not satisfactory. A clear example of the lack of this awareness is the typical response of beginning calculus students (generally the best mathematics students) to the nonroutine problem of calculating the area under the curve $y = x^2 + 2$ between $y = 0$, $x = 0$, and $x = 3$ as illustrated in Figure 1.1 in Chapter 1 (the problem is given before the concept of integration and is the first nonroutine problem they try). Despite directions to determine the area *as best they can*, the great majority of the students calculate the area as 15 1/2 square units (actual area is 15 square units) by "replacing" the curve with the straight lines connecting (0,2) and (1,3), (1,3) and (2,6) and (2,6) and (3,11) and then dividing the area into triangles and rectangles. These students fail to recognize that their calculations could be more accurate by simply increasing the number of divisions (e.g., six divisions yield a solution of 15.125). Notice their solution is inadequate even though the strategy they picked is excellent and, in fact, is the basis of the method of integration in calculus that gives an exact answer to the problem. From having worked with these students, I believe most of the students, typically high achievers, would have been willing to spend the time necessary to persist but simply did not have experience solving problems which require persistence. In practice, approximately 75% of my calculus students determine the best approximation of the area to be 15.5 square units. Of the remaining students, a few even determine the area to be 15 square units (actual area).

In terms of the first step, the students fail to make an initial connection with the solution; i.e., they have no clear sense at the beginning of the problem what an adequate solution (the best approximation of the area) looks like. Notice their solution is inadequate even though the strategy they picked (basically the trapezoid method in calculus) in step two is excellent. In my opinion, the inadequacy of their solution is connected to the first step rather than the third step in the sense that typically a person cannot persist in the sense of step three if there is not an adequate sense of what a good solution might look like (or not look like!) in step 1.

In terms of effectiveness of the curriculum, I have hypothesized (e.g., London, 1993) that if I gave this problem to these students after they completed the curriculum, at least 75% would determine an answer better than 15.5 square units. I have not been able to test that hypothesis with calculus students since by that point in the school year they have learned how to determine the area quickly by integration. However, other data from similar problems requiring the three steps given to calculus students later in the year, and posttest data from the 2004 undergraduate course in which students who did not take a calculus course were give this problem, support that hypothesis.

In conclusion, the three steps of problem solving identified are typically lacking in secondary curricula. In contrast, each nonroutine problem involves these steps and the processing of the problems emphasizes these steps.

Other problem solving skills

Each nonroutine problem gives the student the opportunity to practice the three mentioned steps of problem solving. In addition, selective problems in a sequence of nonroutine problems give the student excellent practice with additional problem solving skills important in mathematics, especially (1) finding patterns and generalizing, (2) developing algorithms or procedures and describing them, (3) manipulating symbols and numbers and (4) reducing a problem to an easier equivalent problem. For example, one problem later in the curriculum for calculus students explores the area bounded by $y = 1/x$, $y = 0$, $x = 1$ and $x = b$ and requires the student to answer three questions – what values of b will yield an area of 1 (see Figure 2.1), 2 and 1,000 (answer: e, e^2 and e^{1000})?

The directions for this problem make it clear that for all practical purposes it would be impossible to determine directly the value of b that would give an area of 1,000; it is necessary to discover a pattern. There are two aspects of this problem that make it particularly valuable as an inductive problem. First, the student is required to generate the data and this is certainly not an easy task. The students are not familiar with the number e and calculating the areas well is a long process. Second, in processing the problem it is emphasized that to be given full credit for an answer of approximately e^{1000} (e.g., 2.7^{1000}) the student must have included direct calculations (or educated estimates) for at least three

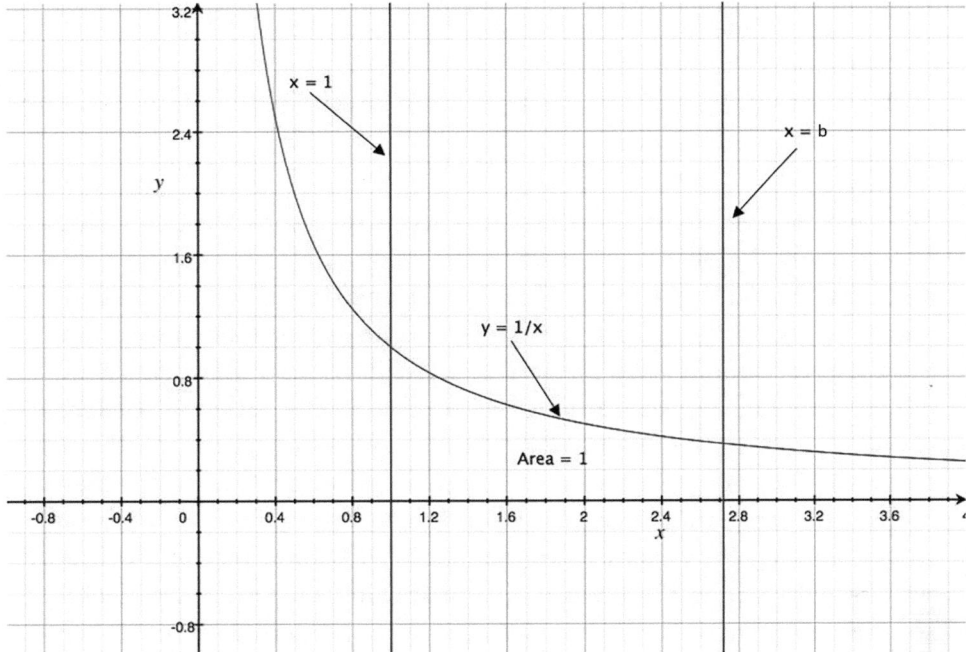

Figure 2.1 The area bounded by y = 1/x, y = 0, x = 1 and x = b.

values of b. For example, if only two values of b are calculated, a student who incorrectly guesses that the pattern is e, e², e⁴, e⁸,… is given the same credit as one who correctly guesses the pattern because both students made the same mistake from a problem solving point of view, generalizing with too little data.

I have observed that students who generalize too quickly on this problem learn from their experience and the processing, and generally do not make that type of error on subsequent nonroutine problems. This indicates something about the value of nonroutine problems. When I have emphasized the need for enough examples to generalize in the context of a lesson in the regular curriculum, many students continue to make that error in lessons in subsequent months. I believe what helps the transfer from the nonroutine problems is the amount of time and effort invested in the solution. After spending two weeks on a problem, the student is more interested in the processing and more likely to retain the substance of the processing.

Mathematical maturity

The best reason I can offer for including a sequence of nonroutine problems in the curriculum is the effect I have observed on the mathematical maturity of the students. Of course, much of this effect is due to the characteristics of nonroutine problems already discussed. However, I believe the whole effect of completing a sequence of nonroutine problems is greater than the sum of the parts. It is as if the student has been transformed mathematically! I have described this transformation as follows (London, 1989):

> The significance of the curriculum can probably be best described by reporting the observed effect on the mathematical maturity of the students. It is as if the student has been transformed mathematically! Instead of acting in all the ways that we normally attribute to most high school students, the student acts similarly to a 'mathematically mature' person. For example, instead of stopping when an obstacle is encountered, the student will persist. Instead of ignoring obvious contradictions or inaccuracies, the student will actively examine them. Instead of being intimidated by ambiguity, the student will tolerate the ambiguity. Instead of being satisfied with the first solution to a problem, the student will work on a problem until a more satisfactory solution is reached. Instead of staring at a problem that seems unsolvable or confusing, the student will try something until the problem naturally becomes clearer.
>
> (p. 1)

These are all examples of the type of behavior that is indicative of the gestalt of a mature mathematician. I am not claiming that merely after completing the nonroutine problems the students are mature mathematicians; certainly, they lack the experience and knowledge of the mathematician and they still make errors "typical" of students but unusual for a mathematician. However, they do begin to act like mathematicians.

A curriculum of nonroutine problems

In this section, I want to give you a sense of what I mean by a curriculum of nonroutine problems by briefly discussing four different ways to integrate a curriculum of nonroutine problems into the mathematics curriculum: (1) a three- or four-year curriculum integrated into a typical content-oriented secondary mathematics curriculum, (2) a typical individual one-year academic mathematics course (e.g., isolated course with no likely follow-up in the next course), (3) a one-year course focusing on the process of problem solving (e.g., a secondary or undergraduate mathematics course) and (4) an integral component of an interdisciplinary curriculum. These approaches will be discussed in more detail in Chapter 10. In addition, I will outline some general considerations in implementing a curriculum of nonroutine problems.

First, I will outline a model for a four-year curriculum of nonroutine problems that can be integrated into a typical high school mathematics curriculum and discuss some guidelines in implementing such a curriculum. I recommend a curriculum consisting of 32 to 60 nonroutine problems (eight to fifteen a year) over the four-year high school curriculum. Each problem typically requiring at least one week to complete (one to three hours of class time) and two-thirds of the problems being solved in cooperative groups. Students would be required to orally and/or in writing document the process for solving each problem. Approximately 40% of the problems would involve content not typically considered mathematical, such as problems that attempt to increase the student's appreciation of diversity, involve ecology problems with an emphasis on problems directly affecting their lives or are practical applications affecting their lives, including problems that students individually define and solve. The curriculum should include an introductory unit that describes the three steps of a nonroutine problem, gives several examples of the steps in a variety of fields, and has the students research additional examples. The curriculum is designed as one component of a secondary mathematics curriculum (less than 20% of the allotted time for mathematics instruction) and is easily integrated with a curriculum consistent with the NCTM standards, as well as with the typical expectations of schools.

In developing the sequence of problems for this curriculum, one major factor was the inclusion of seven strands of content, with a planned four-year sequence of problems within each strand: (1) introductory problems for each year which help establish a good classroom atmosphere; (2) geometry problems; (3) problems involving functions with a focus on understanding abstract concepts; (4) prediction problems; (5) problems that attempt to increase the student's appreciation of diversity; (6) ecology problems with an emphasis on problems directly affecting their lives and (7) practical applications affecting their lives, including a sequence of four problems significantly affecting their lives that each student defines and solves.

In developing a curriculum for one school, we agreed upon requiring a minimum of eight problems a year, with each teacher having a choice of a few problems in a variety of categories. This model satisfied the concern that some teachers had for covering the required curriculum while insuring a reasonable quantity and variety of problems to be effective. In addition, at least a few of the problems were directly related to the content of the course.

In developing a four-year curriculum, attention can be given to gradually shifting the responsibility to students to generate and solve nonroutine problems. For example, it is suggested that the first-year curriculum include an introductory unit (as described previously) in which students study nonroutine problems that others have solved in a variety of fields. In contrast, it is recommended that in the third and fourth year that the curriculum includes a focus on problems generated by the class, as well as a requirement that students individually define and solve significant problems in their own lives.

This model needs to be adjusted slightly if a teacher is teaching a one-year traditional academic course (e.g., Algebra 1) and it is unlikely that the students will have a curriculum of nonroutine problems the following year. In this type of situation, based on my experience and the experience of teachers with whom I have worked, it is difficult to reach the point that students will be able to demonstrate the ability to solve nonroutine problems in their lives. For example, I found that with my work with calculus students, generally they needed at least a sequence of eight problems before they could demonstrate the ability to use the three steps effectively to solve nonroutine mathematical problems, not guaranteeing transfer to other fields. On the other hand, a seventh-grade teacher (DeLeon, 2003) that field-tested a curriculum of ten nonroutine problems for a master's project gathered data that demonstrated that the students were better able to discuss possible solutions to real-life nonroutine problems than a control group. In summary, my experience indicates that even within the context of a one-year traditional mathematics academic course, the students' ability to solve nonroutine problems and do mathematics can be significantly affected; however, you cannot necessarily expect that by the end of the school year the average student would be able to define and solve significant nonroutine problems in their life. In addition to the guidelines already mentioned, I suggest that in such a context you (a) try to include at least 12 nonroutine problems in your curriculum; (b) start with introductory problems (see Chapter 3) that, with support, the students can solve successfully and begin to concretely see the workings of the three steps and (c) include assignments in which you give the students a choice of a few real-life situations involving nonroutine problems and have them describe how they would solve them (e.g., planning an inexpensive and enjoyable vacation). One problem I particularly like in the context of a one-year curriculum is the problem of planning and taking a one-day school trip that is inexpensive, educationally valuable and enjoyable. When appropriately structured, the students are always surprised that they can plan a trip that is much less expensive than the traditional school trip *and* more enjoyable. They concretely see how they can apply the three steps to solve a meaningful problem in a way far beyond their expectations.

A final point needs to be made about the relative role of an emphasis on content and an emphasis on the problem solving skills required to effectively solve nonroutine problems. In developing the curriculum, a variety of measures were taken to minimize the class-time needed to implement the curriculum, but at the same time to insure that enough time was provided to achieve the objectives of the curriculum. Roughly, the curriculum requires less than 20% of the class-time over a four-year period. In addition, some of the problems directly concern mathematical content typically covered in the mathematics curriculum. During field-testing, a full set of nonroutine problems was field-tested in a variety of typical mathematics classes and in those classes we were able to also cover well the traditional curriculum. I find that students actually improve in their ability to do mathematics, making it easier to cover with understanding the content of the academic mathematics curriculum.

The first two models apply to contexts relevant to most teachers; that is, contexts in which the teaching of how to solve nonroutine problems is expected to be integrated into a traditional content-oriented mathematics curriculum. However, in my experience, that type of context de-emphasizes the significance of instruction in the process of problem solving and makes it difficult to maintain a focus on the process. In contrast, I have found that in a one-year course in which the focus is on the process of problem solving, most students demonstrate the ability to solve nonroutine problems involving traditional mathematical contexts and the ability to define and solve nonroutine problems in their own life. For example, in the one-year elective course that I taught, by the second half of the course, students were required to identify and solve at least two nonroutine problems in their life. In addition, the few students that were not seniors and took additional mathematics courses in high school reported improved ability to understand the mathematics in their next course. Therefore, my recommendation for a high school curriculum is that the first year of instruction be a course with the major focus on solving nonroutine problems. I would include problems from the traditional content areas that did not require significant content prerequisite skills as well as at least 40% of the problems from strands not typically associated with the mathematics curriculum. My hypothesis is that such a course would not only teach the process of solving nonroutine problems, but also would prepare students to effectively complete the content typically covered in four years of high school in three years. Philosophically, I need to make two comments. First, theoretically a K-8 mathematics curriculum with an appropriate emphasis on the process of problem solving might make such a course unnecessary. Second, although I directly address how to integrate a curriculum of nonroutine problems in typical content mathematics courses, I believe that there is an inappropriate emphasis on content objectives in many of our present mathematics curricula, implying that we could easily make room for a more in-depth focus on a curriculum of nonroutine problems. In summary, I believe an effective curriculum of nonroutine problems not only improves the students' ability to solve nonroutine problems in mathematics and in daily life, but also improves the students' ability to learn and master new mathematical content.

In addition, I have completed some work in defining alternative models for secondary education (e.g., London, 1996). In that work, one component of the model focuses on the process of problem solving in an integrative manner. Specifically, students take a required course in the process of problem solving which includes a variety of content strands and then apply what they learn in a variety of community contexts (that they generate or select from suggested sites). For example, students can work individually or in small groups with local businesses, nonprofit organizations, local artists and writers and/or local government. Individual work in community contexts would be the basis of discussion in the required core course. A curriculum grid can be developed to insure that students work in a variety of contexts over the course of their education. It should be mentioned that another component of the model suggests a process for identifying core skills required of all students and monitoring their progress in mastering those skills.

Specific suggestions for sequences of nonroutine problems are identified in Chapter 10 for each of the models.

Finally, field-testing indicates that the following guidelines are important when implementing any of the four formats of a curriculum of nonroutine problems:

(1) The most effective way to improve students' ability to solve nonroutine problems is to repeatedly put them in a situation in which they are given a nonroutine problem at an appropriate level of difficulty, have them work on the problem and generate their best solution, and then discuss or process the problem as a class.

(2) In an effective sequence it is normal that each problem *is* difficult. It is normal that the student would not be clear about how to solve the problem initially, at times feel as if the problem is not solvable and need to persist until the problem becomes clear.

(3) The teacher needs to provide a structure that allows students to work on problems at an appropriate level of difficulty. For example, we can manipulate the level of support we provide students when solving the problem. Later I define seven levels of support that form an instructional sequence.

(4) A cooperative group model is consistent with the purposes of this curriculum. For example, students benefit from being exposed to the thinking of other students and cooperative group work provides a supportive atmosphere for dealing with the natural difficulty of the problems.

(5) The curriculum needs to emphasize problems from a variety of fields and problems that are relevant to the student's life. Field-testing indicates that without variety and relevance students are unlikely to transfer the problem solving skills to their day-to-day life. One implication of this guideline is that students with a rich cultural background could benefit from solving problems connected to that background.

(6) Generally, nonroutine problems that do not require significant content prerequisite skills are more effective initial problems than nonroutine problems that require significant content prerequisite skills. Appropriate instruction in solving nonroutine problem does not have to be delayed because of lack of content mastery. For example, I successfully implemented a curriculum of nonroutine problems in a class of ninth-grade students with poor content mastery.

(7) If a teacher feels the need to assign a grade for students' work on nonroutine problems, then a method that primarily evaluates the student's quality of effort is reasonably consistent with this approach.

The fifth guideline, "The curriculum needs to emphasize problems from a variety of fields and problems that directly affect the student's life," is directly connected with the driving question in the development of this curriculum: "What is essential in the mathematics curriculum?" In my opinion, if we are trying to facilitate students understanding what is essential in mathematics, then most mathematics teachers spend much too much time teaching specific content objectives. Far too many students leave secondary and post-secondary schools with the belief that most of what they learned in mathematics is of absolutely no use or value to them in their day-to-day life. On the other hand, I believe if students have an understanding of the three steps of solving a nonroutine problem and how these steps can be used to improve the quality of their lives, then they are learning both the process skills that are essential to doing mathematics well and a way of problem solving that they can use in their day-to-day life. It is clear from the field-testing of the curriculum that to increase the likelihood of transfer to the student's day-to-day life that it is not enough for most students to solve nonroutine problems involving only typical areas of mathematical content. To help insure transfer, the following steps were taken in developing the full curriculum:

(1) A unit was developed that describes the three steps of a nonroutine problem, gives over ten examples of the steps in a variety of non-mathematical fields, has the students research additional examples and requires the students to be able to define and give examples of the three steps.

(2) Approximately 40% of the problems in the curriculum are not typically associated with mathematics and are problems directly affecting the student's lives (e.g., buying an automobile, planning and taking a trip).
(3) Four problems are included in the last two years of the curriculum that require the student to identify and solve a meaningful nonroutine problem in her/his life.

Teaching guidelines

In this section, I will introduce and discuss a variety of guidelines for implementing a curriculum of nonroutine problems, including describing guidelines for adjusting the difficulty level of a nonroutine problem to a level appropriate for your context; suggesting a format for the problem statement for students; discussing the integration of cooperative group activities; identifying appropriate methods for assessment and processing of students' work; providing an emphasis on specific problem solving strategies; and discussing some general considerations in implementing such a curriculum. The teaching guidelines will be further clarified in later chapters that discuss guidelines for specific nonroutine problems, as well as suggestions for a sequence of nonroutine problems for a variety of professional contexts.

Difficulty level

The most effective way to improve students' ability to solve nonroutine problems is to repeatedly put them in a situation in which they are given a nonroutine problem to solve, have them work on the problem for at least a week and generate their best solution and then discuss the problem as a class. For this method to be effective, it is essential that the problems be at the right level of difficulty. The problem has to be "appropriately challenging." Practically speaking, this means that the problem cannot be too easy, or too difficult (e.g., the student finds the problem unproductively frustrating). Typically, the best results occur under the following type of circumstances: the student reads the problem, has no clear idea what the solution is but has a few ideas about how to start, tries something and becomes clearer about the solution, eventually finds a good solution, and, from the processing of the problem, realizes some ways to improve the solution. For any given effective nonroutine problem, typically the problem will be "just right" for perhaps 50% of the students; not challenging enough for 25%; and too difficult for 25%. Of course, teacher scaffolding and challenges can change those percentages.

It needs to be clear that in an effective sequence it is normal that each problem *is* difficult. It is normal that the student would not be clear about how to solve the problem initially; that at times it would feel as if the problem was not solvable; that the first attempts at solution would seemingly lead nowhere; that one would need to persist until the problem eventually became clear. These are normal features of nonroutine problems that students *have to experience* to become good problem solvers! These features differ from unproductive frustration and need to be recognized and encouraged as normal.

At the same time, experience indicates that the great majority of secondary students, without instruction, are not only unable to satisfactorily solve a nonroutine problem, but also would find the experience frustrating or not engaging. Therefore, it is important to provide a structure for the student to move from a point where the process of attempting to solve a nonroutine problem would be a negative experience to a point where they can recognize and define a nonroutine problem, feel motivated to try something and persist until they reach a satisfactory solution. Perhaps the most significant factor affecting the level of difficulty of the problem that we can manipulate is the level of support and assistance we provide the student during the process of solving the problem. I define seven levels of support and assistance that I believe provide a general instructional sequence to help students move effectively to the objective of being able to solve nonroutine problems without significant scaffolding. The levels are:

(1) The teacher defines a nonroutine problem and describes a good solution(s) to the problem. Students are required to summarize the problem and solution.
(2) The students are presented with hypothetical nonroutine problems and are asked to describe the potential obstacles to a solution, a plan or variety of plans to solve the problems and how persistence may be needed to solve the problem.
(3) The students are given a nonroutine problem to solve (either in groups or individually), but the class and teacher brainstorm ideas for each of the three steps before students attempt each step.
(4) The students are given a nonroutine problem to solve (either in groups or individually), and each group or individual brainstorms ideas for each of the three steps for teacher review before attempting each step.
(5) The students are given a nonroutine problem to solve (either in groups or individually) but are required to submit a plan for solving the problem initially and/or a progress report approximately halfway through the process for teacher feedback.

(6) The students are given a nonroutine problem to solve (either in groups or individually) and are not required to use teacher feedback in solving the problem.

(7) The students generate and define nonroutine problems important to them individually (or as a group) and solve each problem.

Within each level, the difficulty of a problem tends to increase as you move from a format involving whole class discussion, to group work, to individual work.

Generally, to achieve the objectives of the curriculum you need to move students through the levels, moving to a higher level when the students have demonstrated a sufficient level of success at the lower level. In practice, the rate at which students move effectively through the seven levels depends on each individual student and class – involving many factors that only the classroom teacher can attempt to judge. The curriculum provides some general guidelines; however, the individual teacher needs to insure that the level for his/her students is both challenging and not unproductively frustrating. Suggestions on moving through the levels include the following:

(1) The curriculum includes an outline in Chapter 11 of an initial unit that moves students through levels 1 and 2. The unit introduces the three steps through (a) describing a variety of examples including the Census Taker Problem and actual examples in a number of different fields, (b) requiring the students to research additional examples and (c) evaluating the student's understanding through an assessment based on their work.

(2) Chapter 3 discusses in detail ten nonroutine problems that are appropriate to introduce the students to the three steps in the process of solving introductory nonroutine problems.

(3) The curriculum includes directions for five problems in the third and fourth year that are problems at levels 6 and 7.

In addition, guidelines are provided throughout the curriculum for insuring that each individual problem is at (or can be adjusted to) an appropriate level of difficulty, as well as suggestions for processing each of the problems. A useful general policy is to let the students know that they can see you individually (or as a working group) when the problem becomes frustrating and that you will give them the assistance necessary to transform the problem to the appropriate level of difficulty. Finally, an important suggestion is that you see the issue of the difficulty of the problems from the students' perspective as a year-long conversation with your class, a conversation that you facilitate whenever it seems appropriate.

Lesson plans for nonroutine problems

There are over 60 nonroutine problems discussed in this book. Therefore, a detailed lesson plan for each problem is beyond this book's scope. In this section, I discuss the major components of such a lesson plan and provide general guidance for each component. Of course, the given lesson guidance is only a guideline that might need to be adjusted to an appropriate level of difficulty for your professional context. The following sections clarify different sections of the lesson plans, providing general guidance appropriate for most contexts:

(1) Goals: Each nonroutine problem is primarily concerned with the goal of improving the students' ability to solve nonroutine problems. In addition, some problems also address specific aspects of this goal and/or specific content goals. If so, these additional goals are identified in the guidelines provided.

(2) Individual versus group problems: Each problem can be completed either individually or in small groups. Approximately 2/3 of the problems are intended to be completed in groups and 1/3 completed individually. From an instructional point of view this distribution seems appropriate. If well organized, group work provides a rich environment for stimulating the improvement of problem solving skills. In addition, (a) students need to learn to work effectively in groups – most real-life problems lend themselves to group problem solving and most jobs require you to work effectively with others, (b) students benefit from being exposed to the strategies and thinking of other students and (c) group work provides a supportive atmosphere for dealing with the natural difficulty of the nonroutine problems. Individual problems give the student an excellent opportunity to internalize and apply what is experienced in the group context. Also, the individual problems give the teacher an opportunity to see more clearly the individual progress for each student.

A given problem is usually more difficult when tackled individually versus by a group. Therefore, if you believe the class or some individuals are not ready for solving a nonroutine problem individually, it would be appropriate to allow them to work in groups (or provide enough hints to convert the problem into one more appropriate). If there is usually a preference for individual or group for a specific nonroutine problem, it will be indicated in the description. Finally, in the beginning of implementation it may be most appropriate to "solve a nonroutine" as a class to help the students to make the transition to more demanding formats.

(3) Written reports and oral presentations: Each problem requires the students to submit a written report documenting the process of solving the problem. The written report provides the basis for organizing the processing of the problem; consequently, from an instructional point of view, the written report needs to clearly and concisely document the process of solving the problem, including all strategies considered (even strategies that did not seem useful), the steps taken to solve the problem and why the steps were taken. My experience is that it takes some time before students can document their work well in the described manner. Therefore, for the first few problems you need to pay particular attention to helping students make the transition to adequately documenting their work. One way to accomplish this is to focus attention on the written reports during the processing of the first few problems. This can be done in a number of ways. For example, (a) in addition to positive remarks concerning their work, you can give back the written reports with specific comments on what needs to be added or clarified and have the groups rewrite the reports; (b) you can select two portions from each report, one positive, one requiring improvement and discuss the reports as a class during the processing; (c) you can take one or a few typical written reports and discuss them in depth as a class or (d) you can require an outline of the written report before it is completed. In any case, an investment of time in the beginning of the curriculum will prove fruitful later in the year.

Each problem solved in groups includes an oral presentation. In addition to giving the students experience communicating orally concerning their work, the oral presentation provides an opportunity to insure that each student in a group understands the group's solution, the steps in arriving at that solution, and why the steps were taken. For example, the teacher can randomly select the person from the group to give the oral presentation, and the presentation can affect 20% of the grade. This provision is usually enough to insure that most students, if not all, will at least understand the group's work (other measures to insure that each student participates in the solution are discussed later). Generally, I allow for two-minute presentations. Usually this is an adequate amount of time, allowing for a fair view of the group's work, yet preserving precious class time.

As was mentioned, each problem provides for an oral presentation and a written report. However, once students are preparing these presentations and reports adequately, it is not necessary to include both for each problem. For example, for some problems, perhaps just an oral presentation is required; for others only a written report. Also, you can omit the provision for an oral presentation with the understanding that you might call on a few students to check for understanding of their group's work.

(4) Cooperative groups: It is beyond the scope of this book to provide comprehensive directions for effectively using cooperative groups. However, since the effective use of groups is an integral part of this approach, I will briefly discuss a few key characteristics of cooperative group work that are integral to this curriculum, as well as providing a few additional references:

a) Group skills: An important aspect of cooperative group work is an emphasis on group skills which helps insure that the quality of the group work supports improvement of the students' skills in solving nonroutine problems. This book does not contain specific instructions concerning the sequencing of group skills or specific group skills to focus on for each problem. It is my experience that the appropriate group skill(s) to focus on naturally presents itself for each individual class of students. Therefore, it would be inappropriate to identify the specific group skill for *your* class to focus on during any given problem. However, I will include here one list of group skills from a handout based on David Johnson and Roger Johnson's work (source unknown, handout at a conference). The list is divided into four levels of skills that provides an excellent framework for planning what skills to focus on with your class. The skills are: Forming Skills: moving into groups quietly, staying with the group, using quiet voices, encouraging everyone to participate, keeping hands and feet to self, looking at the group's paper, using people's names, looking at the speaker and using no put-downs; Functioning Skills: stating and restating the purpose of the assignment, setting or calling attention to time limits, offering procedures on how to most effectively do the task, expressing support and acceptance verbally, expressing support and acceptance non-verbally, asking for help or clarification, offering to explain or clarify, paraphrasing and clarifying other member's contributions, energizing the group with humor, ideas, or enthusiasm and describing feelings when appropriate; Formulating Skills: summarizing the material aloud, seeking accuracy by correcting and/or adding to summaries, seeking elaboration by relating to other learning or knowledge, seeking clever ways of remembering ideas and facts, demanding vocalization of other member's reasoning processes and asking members to plan aloud how to teach material to others and Fermenting Skills: criticizing ideas without criticizing people, differentiating where there is disagreement, integrating different ideas into a single position, asking for justification of others' conclusions or ideas, extending other members' answers or conclusions, probing by asking questions that lead to deeper analysis, generating further answers and testing reality by checking group's work against instructions. You will need to adjust and modify this list to fit your context; however, I believe the sequence provides an excellent framework to adjust to fit your specific needs.

b) The nature of the group work should be interdependent. Some of the group skills stressed above focus on developing interdependence in the group; however, there is certainly an additional need to focus on interdependence in your interactions with the students during the task, as well as in the processing of their work. Some advocates of cooperative groups recommend the use of roles (e.g., recorder, paraphraser, observer) in the group to help foster interdependence. Effective brainstorming also helps to start the group work stressing interdependence. In addition, you can have students work on the problem one night individually before groups are formed and then have each person take a few minutes to share his or her ideas in the group. In general, you need to be aware of the level of interdependence in the groups and take steps to insure a gradual improvement in these skills.

c) Students need to be held accountable for their individual understanding of the group's work. The oral presentations provide the simplest way to check accountability. Some problems include individual transfer items to assess understanding after processing of the problem. In addition, you need to monitor informally the quality of the contribution of each student to his or her group's work.

d) For many problems, the groups are expected to work outside class. Therefore, in forming the groups you need to insure that students are in a group that can meet outside class time, either in person, online or both. If possible, it is useful to provide students with optional access to an online meeting place such as Blackboard or Moodle chat room, Zoom or Google.docs.

e) Varying the composition of groups is useful; however, in the beginning of the curriculum I like to keep students in the same group for a few problems to allow them to develop basic group skills more easily.

f)   The references for this chapter list three excellent books on cooperative groups that are very user friendly. My favorite is Johnson (1993) which gave me the basic information needed to implement effective cooperative groups in my classrooms. Cohen 1986) and Sharan (1992) are also excellent.

(5) Assessment and evaluation: Before suggesting specific suggestions for the evaluation of the students' work on nonroutine problems, it is important to briefly discuss the difference between evaluating content goals and problem solving goals. Typically, when we are evaluating content goals it is appropriate to have students directly demonstrate (e.g., give a test) that they have reached a certain level of mastery. If we are teaching the students to add and subtract fractions with different denominators, at some point we would probably give a test or quiz including problems of that type. If a student failed to meet our criteria (e.g., 80% correct) we usually would be justified in believing that our instruction was not effective for that student. On the other hand, if we had students individually or in groups work on five problems requiring problem solving skills at the appropriate level of difficulty, it would be inappropriate to evaluate the student's learning based on the number of problems answered correctly. The students are likely to improve in their problem solving skills if the problems are at an optimal level of disparity, they make a reasonable effort at trying to solve the problems and the lesson includes a well-constructed processing component. A group that only solves 60% of the problems correctly but worked well on the task and participated in the processing probably would improve more in their problem solving skills than a group that easily answered all the problems! The quality of a student's engagement in the task is a more accurate measure of the likelihood of their improvement. The focus of this book is on the process of solving nonroutine problems, versus specific content objectives. The suggestions for evaluation below try to be consistent with this focus and are implementable in most schools. I will note that in my opinion alternative methods of assessing students (e.g., portfolios) can be much more supportive of student growth than assigning students a number or letter grade; however, I recognize that many teachers do not have a choice concerning the format of evaluation for report cards.

In addition, I believe the experience of working on the nonroutine problems should be a positive experience, suggesting that an approach that creates unnecessary concern for the grade that will be given would be inappropriate. This implies a grading system such as: 90–100% means excellent work, indicating insight and/or effort beyond what one would normally expect; 80–89% means good to very good work, indicating that the student spent a good amount of time on the problem, documented the work and gained reasonable insight into the problem; 70–79% means fair work, indicating that the student put in at least the minimal acceptable effort but clearly could have gone at least a step further; and unacceptable, meaning that the student clearly did not spend enough time to learn anything significant and that the work is incomplete and must be redone. The primary criterion for the grade in this system is the amount of effort the student puts into solving the problem, the quality of the effort primarily affects the grade within each range. For example, an 80 and 85 might indicate the same general level of effort but the 85 would mean that the quality of that student's or group's insight was better. For me, in practice, approximately 10–20% of the students receive a grade in the 90–100% range; 40–60% in the 80–89% range; 20–30% in the 70–79% range; and, at most, one or two students in the unacceptable range. At times, if I feel that the quality of the class's work is not consistent with work that I would judge as adequate, I give back the assignments and allow additional time to work on the problem, giving some hints if appropriate.

In summary, the key concept is that whatever evaluation system you implement, I believe it should be primarily based on the quality of effort, and should not cause unnecessary anxiety for the student, yet demand an acceptable level of effort. In practice, I believe this approach is very fair (as perceived by me and the students) and puts the emphasis on their work versus the grade they receive! In practice, I averaged their grades on the nonroutine problems and entered the average as one test grade. Finally, the uniqueness of each teacher's context requires that I not provide one evaluation rubric for each nonroutine problem. Therefore, I have not included an evaluation rubric for the nonroutine problems discussed in this book. In fact, my method was quite straightforward, I would read through once all the reports to get a general sense of their work as a class, then read them a second time and assign a grade (sometimes more carefully comparing a few). Since my grading was primarily based on effort, I very seldom received a complaint from a student.

(6) Processing: The importance of good processing has been extensively established in research on cooperative groups, as well as more

recently in brain-based research. To help the students, the teacher needs to have an overall sense of the purpose and nature of the sequence of problems. Specifically, the teacher needs to be familiar with the three steps of solving nonroutine problems (problem recognition, trying something and persistence) and other problem solving skills (e.g., finding patterns, and problem reduction). When processing, the teacher should focus attention on these steps and skills whenever possible, especially when a student's solution is a positive or negative example of how to use the steps or skills. The teacher also needs to realize that the typical student requires experience working on eight to ten problems before they can be expected to achieve the type of quality in solutions that we theoretically might want to see. Therefore, we need to have reasonable expectations and patience when processing the problems, especially in the beginning of the school year. It has been my experience that patiently contrasting different student solutions or contrasting teacher prepared solutions with students' solutions is quite beneficial over the course of a school year.

The processing of the problems needs to be carefully prepared, necessarily responding to how *your* students approached the problem. Therefore, I find it helpful in preparing the processing to review each solution and try to pull out selections that will help focus attention on key points. In addition, the teacher can use the processing to draw attention to well-documented solutions, the clever use of resources or unique approaches. Also, the processing can be an opportunity to examine common errors that students make such as not persisting long enough with a problem, generalizing with too little data, carelessness and not exploring conflicting data or obvious errors. After the initial focus on the students' solutions, it is appropriate to then discuss classical or other outstanding solutions and to discuss the historical significance of the problem, some examples of which are identified in the book.

Given the importance of the processing component, I will remind the reader of two representative examples already discussed to clarify the type of processing that is recommended:

(1) As discussed in Chapter 1, one nonroutine problem requires the students to calculate the area under the curve $y = x^2 + 2$ between $y = 0$, $x = 0$ and $x = 3$ (see Figure 1.1). The major focus of the processing for this problem is the fact that most students do not demonstrate a sense of the type of persistence required for an adequate solution. Specifically, despite directions to determine the area "as best they can,"

most students calculate the area as 15 1/2 square units by "replacing" the curve with the straight lines connecting (0,2) and (1,3), (1,3) and (2,6) and (2,6) and (3,11) and then dividing the area into triangles and rectangles (or trapezoids). An "obvious" better solution would be to divide the area into a larger number of trapezoids. It should be added that most students do not divide the area into more trapezoids because they do not have a sense that their solution is not adequate (versus unwillingness to make the necessary effort). Consequently, in processing this portion of the problem, it is important to help the students realize that the strategy of approximating the area by using trapezoids is an excellent strategy, but that the strategy needs to be carried out until an adequate solution is reached. For example, in processing this problem the teacher might first outline the typical solution of 15 1/2 square units. Then, if any students used this strategy with more than three trapezoids, their solutions could be discussed, emphasizing the increased accuracy and any stated student realization of the need to increase the number of trapezoids. Here are some examples of quotes from student solutions that would be appropriate:

> "This solution [15 1/2 square units] seemed like it could be improved. When I blew up the graph I could see that I was over on my answer.... I decided to divide the graph by thirds [i.e., nine trapezoids]."
> "At a certain point I saw that the area was approaching 15. I would increase the number of divisions and get basically the same answer.... So my answer is 15."
> "But I wasn't satisfied [with the solution of 15 ½]. So I pressed on by trying more divisions…"

The emphasis in this portion of the processing is to make the students aware of the affective component necessary for an adequate solution - dissatisfaction with the initial solution.

(2) One advanced nonroutine problem requires the student to figure out the values of t such that the area bounded by $y = 1/x$, $y = 0$, $x = 1$ and $x = t$ is 1, 2 and 1,000 (see Figure 2.1). The actual solutions are $t = e^1$, $t = e^2$ and $t = e^{1000}$. A key concept in the processing is the proper use of inductive reasoning. For example, many students will generalize an answer for the value of t for an area of 1,000 based on just two bits of data (areas of 1 and 2), not realizing the need for more examples. One method to emphasize this need in the processing is to have

the students compare the following solutions from three students in one class (for consistency, we will assume that each student found t = 2.71 for an area of 1 and t = 7.37 for an area of 2):

The first student

noticed that the difference between t for an area of 1 and t for an area of 2 is 4.66 which is about 4 2/3…. So in order to get the value of t for an area of 1000 you must multiply 999 times 4 2/3 and add it to 2.71. [Therefore,] the value of t for an area of 1000 is 2.71 × (4 2/3)(999).

The second student

found that when you square the t value for the area equal to 1 the answer is very close to the value of t when the area equals 2. I therefore concluded that when you put the value of t [when the area equals 1] to the power of the area, you will find the corresponding t value for the area equal to 1000 [t = 2.71^{1000}].

The third student

decided to take the square root of t when the area under the curve equals 2…. That number is surprisingly close to 2.71, therefore I assert that t equals 2.71 raised to a power equivalent to the desired area under the curve. I tried this formula with areas of a half and 0 and it seemed to hold. [Therefore, I think] t would equal 2.71^{1000} when the area under the curve equals 1000.

After student discussion, the teacher should point out that the first and second student made the same reasoning error: generalizing with too few examples; therefore, from a problem solving point of view the two students should receive the same grade, even though one student obtained the "correct" answer. In contrast, the third student formed a hypothesis and tested it with two additional values. I will note that in practice, all else being equal, I would give the first two students a grade in the 80–89 range, and the third student a grade in the 90s.

The six above components of a lesson on nonroutine problems need to be considered in planning every lesson on nonroutine lessons. In addition, for some problems the following two components might be important:

(1) Progress reports. The progress report is just a simple summary of what has been done so far. It allows the teacher not only to check

that the student or the group has been working on the assignment, but also gives the teacher the opportunity to give assistance or hints to a few appropriate students or groups. In addition to when the progress report is suggested in the directions, a progress report is appropriate whenever you believe it is needed to insure a reasonable level of quality or you believe that additional scaffolding may be appropriate. Also, you might assign a progress report for certain groups or individuals whose work has not been as consistent as other students' work.

(2) Extensions and enrichments. After the initial focus on the students' solutions, it is appropriate to then discuss classical or other outstanding solutions and to discuss the historical significance of the problem. Also, extensions of the problems have been suggested when appropriate. Obviously, each teacher has to decide how much time, if any, is available for any given problem for enrichment or extensions. Typically, due to time restraints, the teacher will have to pick the extensions that seem most likely to interest his or her students. There are a number of formats for enrichment or extensions that should be considered, including: whole class discussions, extra credit assignments, a second related nonroutine problem completed individually or in groups or a requirement that students select a certain number of enrichment activities or extensions to complete in a given period of time.

A curriculum of nonroutine problems: some considerations

A curriculum of nonroutine problems is not like the traditional mathematics curriculum and cannot be effectively taught with the same expectations. At a certain point in the curriculum, most students feel comfortable with the three steps of a nonroutine problem and even feel a sense of adventure when given the next problem. In contrast, in the beginning of the curriculum, many students feel uncomfortable with the three steps and wish that the problems were "easier" or not part of the curriculum at all. I do not believe that these initial feelings are natural, but rather the result of instructional practices over many years that prevent students from experiencing the three steps, and, in fact, encourage behaviors contradictory to what is required to solve a nonroutine problem. The curriculum of nonroutine problems is a *gradual* process of helping the student feel comfortable with the process of solving a nonroutine problem. For the curriculum to work, it is essential that the teacher understand this aim of the curriculum and feel comfortable supporting the students through the gradual process of changing the way they view mathematics and how they solve problems. I can guarantee that the required patience and gentle insistence on the aim of the curriculum will reap a bountiful harvest!

For the first nonroutine problems, it would be wise to spend some time discussing expectations. For example, you might discuss the following points:

(1) The problems require at least a few hours work spread over the period of one to three weeks – they will not be able to answer the problems adequately if they save the assignment for the last day or two

(2) Part of the assignment is to document all nontrivial strategies and solutions attempted as well as the final solution, including justification for selecting the final strategy.

(3) There is not one "best" solution for these problems, but rather a variety of good solutions, including some novel solutions.

(4) It is quite common with a nonroutine problem that they will "solve" the problem one way and then realize a method to refine the solution. Point out that you encourage this and want them to document it as part of the assignment.

There are probably a variety of ways to effectively structure the assignments. A structure that has worked well for me is to assign one problem every two to three weeks. The problem is introduced in part of one class and typically collected two weeks later. On occasion, I will give the students a week or two off either to allow for a more intense focus on the content of the course or just to give them a break from nonroutine problems. In giving the specific directions for each problem, it is important to give the students enough information so that they can engage the problem but, at the same time, not give them so much information that the problem becomes straightforward. If the directions are good, the solution will not be straightforward yet each student will be able to work on the problem without unproductive frustration and will be able to generate a solution. In summary, the problem will be challenging yet not impossible.

The teacher should encourage the students to work at least some on the problem the first night, emphasizing that this type of start allows ideas to develop even when the student is not directly working on the problem. Also, I like to communicate to students that they are encouraged to use other resources. For example, for each problem I like to make sure that the students realize I am available as a sounding board for ideas. When students do approach me for assistance I certainly do not tell them how to solve the problem but rather try to give the students feedback or hints that will allow them to work productively. Also, they are encouraged to use outside references and resources when appropriate such as computers, science equipment, and math books from previous courses. Of course, a distinction is made between using a resource as an aid in solving a problem and using a resource to solve a problem. It is inappropriate to ask a mathematician for a solution or to try to look up a solution in a book.

The common core mathematical practices

What is the connection between the CCMP and a curriculum of nonroutine problems? The purpose of this section is to outline what I believe to be the consistency between the approach to instruction for a curriculum of nonroutine problems and the implications for instruction of the CCMP, and to outline in some detail the specifics of how a curriculum of nonroutine problems can facilitate development in the students of the CCMP.

First, I need to clarify that the pedagogy described for a curriculum of nonroutine problems was developed before the CCMP were adopted; that is, the pedagogy described was not developed in response to the adoption of the CCMP. However, my approach to pedagogy as well as the approach of many leaders in the field of mathematics education has been significantly affected by the emphasis for many years on improving the problem solving skills of mathematics students (e.g., the NCTM standards), which emphasis significantly affected the development of the CCMP. Basically, a curriculum of nonroutine problems is one of a variety of approaches to mathematics education prior to the formal definition of the practices that facilitates the improvement of students' ability to do mathematics, including their ability to integrate the CCMP into their work in mathematics classes, as well as in their daily life.

Before suggesting a framework for integrating an emphasis on the CCMP, I will note that in my opinion the framework for the nonroutine problems is more holistic and less discrete than the CCMP. For example, the CCMP suggests eight different practices that though there are obvious connections between some of them they are described as eight different practices to be nurtured, perhaps individually, a few together, or even all eight in a project. In contrast, by definition, each nonroutine problem requires the same three steps to solve and processing is organized around the whole picture of the three steps and how the steps were implemented in the process of solving the problems. Of course, some problems lend themselves to emphasizing one step over the others but the vision is clarifying the whole picture/process. Also, in general mathematics teaching (versus a curriculum of nonroutine problems), it would not be unusual that the processing questions for the CCMP would be determined prior to the completion of students' work;

e.g., a specific question that focuses on what tools they used to solve the problem and reflecting on whether the tools were effective. For nonroutine problems, while there are some general questions that are suggested for processing, the most significant questions naturally emerge from the students' solutions and the processing of their work.

Theoretically, a curriculum of nonroutine problems promotes all the CCMP; however, in practice some are emphasized in all nonroutine problems, and some more emphasized in certain problems. In addition, each CCMP can be addressed in at least a few different ways that may differ significantly from one another. Finally, since the nonroutine problems emphasize an open-ended approach to strategies to solve a problem, which mathematical practice it is appropriate to emphasize is at least somewhat dependent on the strategies the students select to attempt.

Due to space limitations, I will discuss just two of the CCMP and how a curriculum of nonroutine problems provides a natural context to deeply focus on the CCMP. I will limit my examples of nonroutine problems in this section to problems that are clearly mathematical in nature. Finally, I will try to summarize some implications for the approach discussed for improving students' ability to apply the CCMP in their mathematics education as well as their life. I will explore the first common practice, "Make sense of problems and persevere in solving them," primarily through just one example to emphasize the connection and relevance between the first practice and *every* nonroutine problem, as well as illustrating the lack of this practice in even advanced mathematics students, and the implications for instruction. Also, I will discuss one additional practice, with two problems, one appropriate as an introductory nonroutine problem, and one appropriate later in the curriculum.

More than the other CCMP, common practice 1, "Make sense of problems and persevere in solving them," is most consistent with the definition of a nonroutine problem and is therefore a common practice addressed in every nonroutine problem. Specifically, "Make sense of problems" corresponds well with the first step of solving a nonroutine problem, "problem recognition," and "persevere in solving them" corresponds well with the third step of solving a nonroutine problem, "persistence." The second step of solving a nonroutine problem, "trying something" is certainly assumed or implied in the wording of the first common practice.

The primary example I want to use for this practice not only clarifies the step of persisting or persevering in solving a problem, but also illustrates that even most of the best secondary mathematics students have difficulty (without instruction) with this step. The step of persistence implies having a connection with what a satisfactory solution looks like. For example, persistence implies that the problem solver realizes that a particular solution is not satisfactory. A clear example of the lack of this awareness is the typical response of beginning AP Calculus students (generally the top 10% of the mathematics students) to the nonroutine problem of calculating the area under the curve $y = x^2 + 2$ between $y = 0$, $x = 0$ and $x = 3$ as illustrated in Chapter 1 and earlier in this chapter (Figure 1.1). The problem is given before the concept of integration and is the first nonroutine problem they attempt to solve. Despite directions to determine the area *as best they can*, the great majority of the students calculate the area as 15 1/2 square units (actual area is 15 square units) by "replacing" the curve with the straight lines connecting (0,2) and (1,3), (1,3) and (2,6) and (2,6) and (3,11) and then dividing the area into triangles and rectangles.

These students fail to recognize that their calculations could be more accurate by simply increasing the number of divisions (e.g., nine divisions yield a solution of 15.125). Notice their solution is inadequate even though the strategy they picked is excellent and, in fact, is the basis of the method of integration in calculus that gives an exact answer to the problem. From having worked with these students, I believe most of the students, typically high achievers, would have been willing to spend the time necessary to persist but simply did not have experience solving problems which require them to recognize that a problem does not always have a straightforward solution. Certainly typical textbook problems do not require this skill. In practice, approximately 75% of my calculus students determine the best approximation of the area to be 15.5 square units. Of the remaining students, a few even determine the area to be 15 square units (actual area).

I will also note that in terms of the first step, problem recognition, the students fail to make an initial connection with the solution; i.e., they do not have a clear sense at the beginning of the problem what an adequate solution (the best approximation of the area) looks like. Notice their solution is inadequate even though the strategy they picked (basically the trapezoid method in calculus) in step two is excellent. In fact, it could be argued that the inadequacy of their solution is connected to the first step (e.g., a lack of connection with step three; i.e., a sense of what a good solution might look like and some sense of the obstacles to a solution) rather than the third step in the sense that typically a person will not persist in the sense of step three if there is not an adequate sense of what a good solution might look like in step 1.

This example provides a good opportunity to discuss one approach to improving the students' ability to effectively implement the CCMP. Specifically, in working with these students I have found that after a curriculum of eight nonroutine problems over three quarters of a school year, integrated into a calculus course, the students demonstrated a good mastery of this common practice. I have hypothesized that if I gave this area problem to calculus students after they completed the curriculum, at least 75% would determine an answer better than 15.5 square units (versus 25%). I have not been able to test that hypothesis directly with calculus students since by that point in the school year they have learned how to determine the exact area quickly by integration. However, other data supports that hypothesis: (1) data from similar problems requiring the three steps given to calculus students later in the year and (2) data from a ten-week undergraduate liberal studies problem solving course (non-mathematics majors) involving a curriculum of nonrou-tine problems in which students who did not take a calculus course were give this problem as part of a posttest.

In conclusion, the three steps of problem solving identified are typically lacking in the secondary mathematics curriculum. In contrast, each nonrou-tine problem involves these steps and the processing of the problems empha-sizes these steps.

The fifth common practice, "Use appropriate tools strategically," includes the ability to use technological tools to explore and deepen the student's understanding of a problem, and the ability to consider the available tools to help solve a problem and make sound decisions about when a tool might be appropriate, as well as recognizing the limitations of the tool. I will illustrate how this common practice can be addressed in a substantial way in the con-text of a curriculum of nonroutine problems by discussing two nonroutine problems that require technology as a component for solving the problems, but also require the student to "make sound decisions about when a tool might be appropriate, as well as recognizing the limitations of the tool." One problem will be appropriate for students with experience solving nonroutine problems and one will be appropriate as an introductory problem in a curric-ulum of nonroutine problems.

First, the problem, "Y = sin x − e^x," is appropriate for more advanced students with experience solving nonroutine problems. The problem asks students to estimate as accurately as possible the 50th root of y = sin x − e^x (e ≈ 2.714 and is an important number in calculus) to the left of x = 0 (y-axis). You cannot solve the equation for x through normal algebraic manipulation, rather the student needs to generate data using a graphing calculator and look for patterns and persist until they are convinced they have reached an excellent solution. In working with calculus students with experience solving nonroutine problems, a typical good student solution is as follows: (a) Prob-lem recognition: the student realizes the problem is not solvable by algebraic manipulation, sees the need to generate data and has some confidence that trying something and persisting will lead to a solution; (b) trying something: the student uses a graphing calculator to generate the first few roots (the first three roots are approximately -3.1830, -6.2813 and -9.4249) and looks for patterns, noticing a fairly linear decrease in the value of the roots. For example, some students plot the points of the form: (the number of the root, the value of the root), and fit a straight line to the points and notice that the slope is approximately π (or perhaps initially they believe it is a slope of 3) and (c) persistence: in trying to make sense of the pattern the student realizes that the roots change by a value approaching π ($\pi \approx 3.14$) and realizes that e^x contributes close to zero to the value of $\sin x - e^x$ as x decreases in value (absolute value of x increases); therefore, the appropriate root of $y = -\sin x$ (i.e., $x = -50\pi$) is an excellent approximation of the actual root. In fact, when I have tried this problem with calculus students with experience solving nonroutine problems, the majority of the students found an excellent solution (versus less than 25% without similar instruction). Not only is their answer excellent, but they understand why it is excellent. In contrast, students who try to solve this problem by using the calculator to literally find the 50th root either make an error in the process or get an approximation limited by the rounding off process of the calculator. These students are using technology, but not wisely and not "recognizing the limitations of the tool." For example, they more than likely have no understanding of why their answer is appropriate (e.g., when e^x approaches 0, the appropriate root of $-\sin x$ is an excellent answer) and they do not realize that manually solving it with a graphing calculator is likely to result in errors due to the limitations of this tool for this problem. In other words, this is an example of a nonroutine problem that naturally requires technology, but also requires students to use higher order thinking skills such as common practice 5 to reach an excellent solution.

The second problem, "Five Calculations" is a nonroutine problem that is particularly appropriate as an introductory problem with scaffolding (see Chapter 3). The problem requires the use of calculators and the ability to make sound decisions concerning how and when the tool might be appropriate, as well as recognizing the limitations of the tool. For this problem, each group of three students prepares for the following task: For each of 20 numbers, between 100 and 900, provided by the teacher (i.e., the specific numbers are unknown to the students before the task), reduce each number to 0 with five or fewer calculations involving addition, subtraction, multiplication and/or division by the whole numbers 1 to 9 only. For example, given the number 417, one solution would be: (a) subtract 1: $417-1 = 416$, (b) divide by 8: $416/8 = 52$, (c) add 4: $52 + 4 = 56$, (d) divide by 7: $56/7 = 8$, and (e) subtract 8: $8-8 = 0$. For the task assessment, each group is given two calculators and 20 minutes to complete the task. In addition to blank paper and writing utensils, each group may bring one 3 by 5 card with writing on one side. Groups are given one point if they reduce a given number to 0 in five calculations; two points if they reduce a given number to 0 in four calculations; and three points if they reduce a given number to 0 in three calculations. The groups have one to two weeks to prepare for the task (primarily outside class).

A discussion of one solution will clarify the nature of the problem. It turns out that there is a fairly straightforward method to generate a solution in five or fewer steps for the numbers 100–819: (a) if not divisible by 9, subtract a number less than 9 to get a multiple of 9, (b) divide by 9 to get a number less than 91, (c) if not divisible by 9, subtract a number less than 9 to get a multiple of 9, (d) divide by 9 to get a number less than 10, and (e) subtract to get 0. For example, if given 784, then $784 - 1 = 783$; $783/9 = 87$; $87 - 6 = 81$; $81/9 = 9$; and $9 - 9 = 0$. The numbers from 811 to 900 are not as easy. One effective method requires you to find numbers in the 811–900 range that allow two consecutive divisions by fairly large factors (less than or equal to 9 of course) or represent other similar special circumstances that allow reduction in four steps. Then you can generate an interval of solutions in five steps by first adding or subtracting to get that number. For example, the numbers from 855 to 873 can be reduced to 0 by first getting to 864 by adding or subtracting, then $864/9 = 96$, $96/8 = 12$, $12/6 = 2$, and $2 - 2 = 0$. It turns out that by this process all numbers except 850 to 854 can be reduced in five steps. Finally, 850, 852 and 854 can be reduced by individual solutions (e.g., $854/7 = 122$, $122 - 2 = 120$, $120/6 = 20$, $20/5 = 4$, and $4 - 4 = 0$). There is no solution in five steps for 851 and 853. If a group discovers the solution for the numbers 100–810, there is plenty of room on the 3 by 5 card to write the solutions (using intervals for most) for 811–900. Investigations of situations that can be solved in less than five steps and consideration of how to work effectively as a group with limited resources and time will further improve a team's score and allow an opportunity for persistence.

The wise use of technology (calculators) is particularly essential during the timed task. For example, while a solution (or lack of solution) in exactly five steps can be straightforward during the task based on the above solution and notes on your 3 by 5 card, finding solutions in 3 or 4 steps generally need to be found during the actual task. For example, a reduction in three steps requires (minimally) that your first two steps are divisions. A reduction in four steps requires (minimally) that at least two steps of your first three steps are divisions. While your notes can help in this process (e.g., listing multiples of three factors), to get the best score on the task requires well-coordinated and effective use of the two calculators.

From the point of view of mastering nonroutine problems, this problem is excellent for early in the curriculum. The problem requires few content prerequisite skills and no need to use written or people resources outside the group. In addition, the problem lends itself naturally to a higher level of support by giving the groups one or two field-tests before the final test – a nice way to provide a transition from trying something to persisting. For example, in practice, most groups in an average academic ninth or tenth grade class believe they have developed a good strategy before the first field-test and wonder why a field-test is necessary and they are surprised to find that their strategies are not nearly as effective as expected! This problem is equally appropriate for more advanced students. For example, I have field-tested the problem with calculus students and they also overestimate the effectiveness of their strategy for the first field-test (generally not as much as average students). The structure of requiring the field-test provides the necessary support to help the students focus on the need for persistence in this problem. My hypothesis is that later in the curriculum the students would have naturally field-tested their strategy without the need of the teacher requiring or suggesting it.

Finally, I will mention that I do not include 851 or 853 in the first field-test, but include one of them in the final field-test, a good measure of if they persisted in an outstanding manner and a good basis for the processing related to persistence.

Notice that this problem deals with content not considered particularly important in the academic high school curriculum. So, from the point of view of mastering content this task would not be considered very important; however, from the point of view of mastering nonroutine problems and the CCMP I have found this task to be excellent, particularly for the first year of the curriculum.

Implications for improving the CCMP

The previous section only briefly explores an approach to improving the CCMP as a component of a curriculum of nonroutine problems. However, based on the given examples and extensive field-testing, I would like to summarize what I see as the implications of the outlined approach for improving the CCMP in the context of a curriculum of nonroutine problems. Of course, I am only describing what I consider one component of a more comprehensive approach that would be needed to fully address the CCMP.

Three particularly relevant guidelines mentioned earlier in my discussions concerning pedagogy for a curriculum of nonroutine problems are relevant to the CCMP and can be briefly summarized:

(1) One effective approach to the CCMP is to repeatedly put the students in a situation in which they are given a significant problem requiring a good investment of time (e.g., a nonroutine problem) at

an appropriate level of difficulty, have them work on the problem and generate their best solution and then discuss or process the problem as a class.

(2) In an effective sequence of nonroutine problems it is normal that each problem is difficult and that the student would not be clear about how to solve the problem initially, at times feel as if the problem is not solvable and need to persist until the problem becomes clear. A cooperative group approach for a majority of the problems helps address this natural difficulty in a supportive manner.

(3) The teacher needs to provide a structure that allows students to work on problems at an appropriate level of difficulty. For example, instruction needs to gradually move students from requiring significant scaffolding to solving problems requiring no or minimal scaffolding.

Two additional principles that were not emphasized in the previous section, but are essential: First, for many students, instruction in the process of problem solving (e.g., the CCMP), at least initially, tends to be most effective when not directly integrated into the teaching of mathematics content. For example, instruction for which the major objective is mastery of a specific content objective is not recommended as the only vehicle for improvement of the CCMP. In my experience the academic achievement of my students has consistently been as much as, if not more, than other comparable sections. This may seem counterintuitive; however, effective instruction focusing on the process of problem solving naturally improves the student's ability to learn and to do mathematics. In addition, such instruction can be integrated in a way that minimizes classroom instructional time.

Second, instruction in the process of problem solving and the CCMP must insure that students attempt meaningful problems in a variety of fields to facilitate transference of the skills to other fields and daily life. My initial field-testing with calculus students did not integrate these two principles. For example, their relative comfortableness with mathematics content minimized the need for concern about content prerequisite skills and anxiety. When I implemented this approach with average and below average ability mathematics students, the significance of the need to minimize the connection to the mathematics concepts in at least the initial phases of the curriculum became obvious. The need to integrate a variety of problems from fields not generally considered mathematical only became apparent after the first few years of field-testing, especially in the context of a year elective high school course in problem solving and an undergraduate problem solving course. Specifically, after a few years of field-testing with the problem solving class, I integrated into the four-year curriculum of nonroutine problems a few strands emphasizing "nonmathematical" problems such as problems which attempt to increase the student's appreciation of diversity, especially diversity of ethnic groups and cultures; ecology problems with an emphasis on problems directly affecting their lives and practical applications affecting their lives, including a sequence of four problems (starting in the third year) affecting their lives that each student defines and solves. These strands accounted for approximately 40% of the problems in the curriculum. Posttest data supports the hypothesis that the students were able to apply the problem solving skills to meaningful "nonmathematical" problems. Many mathematics educators have emphasized the importance of integrating instruction in the CCMP with instruction in the content standards – the above two points in my opinion are not inconsistent with the intent of that emphasis; that is, from my experience these two approaches increase the ability of the mathematics educator to *effectively* integrate the understanding of the content with an emphasis on the CCMP. In addition, a focus only on "mathematics content" reduces the likelihood of transfer to other types of content.

Finally, I want to reiterate some of the key assumptions embedded in the examples and implied by the fact that the approach to a curriculum of nonroutine problems is comprehensive, holistic and consistent with a constructivist approach:

(1) A curriculum of nonroutine problems is holistic in the sense that, by definition, each nonroutine problem requires the same three steps to solve and processing is organized around the whole picture of the three steps and how the steps were implemented in the process of solving each problem. Similarly, it is suggested that activities that are meant to improve the CCMP need to be primarily organized around the whole picture of the direction of the CCMP versus discrete questions for each of the eight practices.

(2) The focus of the processing for each problem should naturally emerge from the students' solutions and the processing of their work; that is, while there may be some general questions that are suggested for processing most problems, the most significant questions or focuses naturally emerge from the students' solutions and the processing of their work.

(3) The process of improving students' ability in the CCMP and in solving nonroutine problems should be framed as an ongoing conversation with the students and can include addressing the fact that each problem is difficult and helping students become comfortable with feelings generated by that difficulty, helping the students notice improvements

in their abilities and discussing what contributed to the improvement, noticing patterns of difficulty and discussing how to address the difficulties and helping them realize that improvement will *gradually* occur over the course of the year. In summary, this type of approach to improving the CCMP helps facilitate the type of coherent focus that is needed to significantly improve students' ability to integrate the CCMP into their problem solving process both in mathematics education and in solving significant real life problems.

Concluding remarks

In this chapter, I have attempted to give the reader a sense of what is meant by a nonroutine problem and a curriculum involving a sequence of nonroutine problems. Hopefully, I have conveyed my enthusiasm for this curriculum. I believe that a sequence of nonroutine problems has a richness to it that allows the student to reach a level of mathematical maturity not typically achieved in other approaches.

Obviously, the sequence is more than 60 separate problems. In this chapter, I have tried to provide the theory and suggestions necessary for the reader to see the connections among the problems, thereby facilitating the creation of a learning environment conducive to mastering the objectives of the sequence. I have tried to be complete in the suggestions in this chapter and the remainder of the book, but one essential suggestion has been omitted, allow yourself to be open to the richness of the problems, to the variety of solutions and to the opportunities for growth that the student efforts will create.

References

Brooks, J. and Brooks, M. (1993). *In search of understanding: The case for constructivist classrooms*. Alexandria, VA: Association for Supervision and Curriculum Development.

Cohen, E. (1986). *Designing group work*. Teachers College Press, New York.

Curtis, D. (2002). The power of projects. *Educational Leadership, 60*(1), 50–3.

DeLeon, A. (2003). *A curriculum of nonroutine problems in the middle school*. Draft of master's project. California State University, San Bernardino.

Hickey, D., Moore, A. and Pellegrino, J. (2001). The motivational and academic consequences of elementary environments: Do constructivist innovations and reforms make a difference? *American Educational Research Journal 38*(3), 611–52.

Johnson, D., Johnson, R. and Johnson, E. (1993). *Circles of learning: Cooperation in the classroom*. Holubec, MN (Minnesota). Interaction Book Company.

Krulik, S. and Rudnick, J. (1980). *Problem solving: A handbook for teachers*. Boston, MA: Allyn & Bacon.

London, R. (1976, April). *Process-oriented curriculum*. Paper presented at the American Educational Research Association National Convention, San Francisco.

London, R. (1993, April). *A curriculum of open-ended nonroutine problems*. Paper presented at the American Educational Research Association National Convention, New York.

London, R. (2004). What is essential in mathematics education? A holistic perspective. *Encounter: Education for meaning and social justice 17*(2), 10–17.

London, R. (2007). What is essential in mathematics? MSOR connections, 7(1), 10-17.

March, T. (2004). The learning power of WebQuests. *Educational Leadership, 61*(4), 42–7.

Moulds, P. (2004). Rich tasks. *Educational Leadership, 61*(4), 75–8.

Peressini, D. and Knuth, E. (2000). The role of tasks in developing communities of mathematical inquiry. *Teaching Children Mathematics, 6*(6), 391–7.

Polya, G. (1962). *Mathematical discovery: Volume 1*. New York: John Wiley & Sons.

Schoenfeld, A. H. (1985). *Mathematical problem solving*. Orlando, FL: Academic Press.

Sharan, Y. and Sharan, S. (1992). *Expanding cooperative learning through group investigation*. New York: Teachers College Press.

Sobel, D. (2005). *Place-based education: Connecting classrooms and communities*. New York: The Orion Society.

Ward, C. (2001). Under construction: On becoming a constructivist in view of the Standards. *Mathematics Teacher, 94*(2), 94–6.

3

Introductory problems

Most students have little or no experience solving nonroutine problems; that is, problems that require the students to try something even though it is not clear or likely to lead directly to a solution and to persist in trying to solve the problem until a good solution is found. In this chapter, I will discuss a number of problems that introduce the students to the three steps of nonroutine problems in a way consistent with their initial skills, thereby helping them make the transition to solving more difficult nonroutine problems with less scaffolding than these introductory problems. I identified the need for these introductory problems based on my experience working directly with students, I was surprised many times by their lack of understanding of the process of solving a nonroutine problem, even my calculus students! Specifically, in this chapter I will discuss in some detail ten nonroutine problems involving mathematical content that I have found to be very useful in facilitating students experientially becoming comfortable with the three steps of solving a nonroutine problem, thereby providing the foundation for the students to effectively work on additional nonroutine problems with less scaffolding. First, I will discuss in depth the "Expensive tape" problem not only with the intention of giving the reader an understanding of the problem, but also as an opportunity to concretely illustrate some of the principles of pedagogy discussed in the introductory chapters. Second, I will discuss four additional nonroutine problems that can be completed in class (perhaps requiring a few class periods), providing a reasonably controlled environment to insure the appropriate level of challenge and to provide an excellent environment for

DOI: 10.4324/9781003393283-3

processing and feedback. Finally, I will discuss five additional introductory nonroutine problems that provide a transition to problems requiring individual or group work outside of class. Of course, you need to reflect on how to implement the curriculum in a way consistent with the needs of your students and your philosophy and comfort as a teacher, including how many, if any, of the ten would be appropriate for your context. Hopefully, these chapters will provide a basis for you to make an individual decision appropriate for your context!

Expensive Tape Problem

The Expensive Tape Problem is one of my favorite problems for introducing the three steps of a nonroutine problem to students in the context of a reasonably meaningful mathematics problem (versus the Census Taker Problem) that can be completed in one or two classroom periods with appropriate scaffolding. Structured properly, students experientially see the power of trying something to solve a problem that appears beyond their capabilities to solve. In addition, most of the time, the processing of the problem clearly introduces the students to the third step of persistence.

A brief statement of the Expensive Tape Problem: Your group is a research team for a company. You are given the following problem: The company sends out many packages and the company wants to spend less money on packaging. By far the most expensive cost for packaging is the special tape needed to properly secure the box. When wrapping the box, you need to tape one length and the girth (around the middle of the box – two widths and two heights). You are required to: (a) use exactly 30″ of tape per box, (b) have at least two of the three dimensions (length, width, and height) be a whole number of inches and (c) create a box that can hold the most material. The company wants your three best designs for maximizing the volume for a box.

Before discussing the suggested scaffolding for this problem, I will discuss the solution to give you a context for understanding the rationale for the scaffolding. If you generate some data, you should soon begin to notice two patterns that make the problem easier. Below is some data that illustrates the two patterns:

The two patterns illustrated are (a) girths closest to squares (i.e., the width and height are equal or close to equal) yield higher volumes and (b) lengths closer to ten inches tend to yield higher volumes. Mathematically, the first pattern follows from the fact that the rectangle with the maximum area for a given perimeter is a square. The second pattern follows from the solution of

Length	Width	Height	Volume
10	5	5	250
10	6	4	240
10	7	3	210
10	8	2	160
10	9	1	90
16	4	3	192
14	4	4	224
12	5	4	240
10	5	5	250
8	6	5	240
6	6	6	216
4	7	6	168

a calculus problem maximizing the volume of the box with the given restrictions (of course, beyond the skills of the typical student). Given time to gather data, most groups notice these two patterns in the data, thereby finding good solutions to the problem, such as the following three boxes: 10″ by 5″ by 5″ (250 cubic inches); 8″ by 6″ by 5″ (240 cubic inches); and 12″ by 5″ by 4″ (240 cubic inches). Notice that these three solutions are consistent with the two patterns: the lengths are close to 10″ and the girths are as close to squares as possible with the restrictions. Students that persist realize that 9″ and 11″ lengths should in theory provide better solutions than 8″ and 12″ lengths (being closer to 10″), and discover the following two solutions, each of which includes one fraction: 9″ by 5″ by 5 1/2″ and 11″ by 5″ by 4 1/2″ (both volumes equaling 247.5 cubic inches). It should be noted that many students avoid lengths that are odd numbers because they assume (incorrectly) that odd lengths will violate the restriction concerning "at least two of the three dimensions (length, width and height) must be a whole number of inches" (see discussion below).

The Expensive Tape Problem is the nonroutine problem I have used most often to experientially introduce students to the three steps of a nonroutine problem. As a result, I have refined the directions and scaffolding for students to help insure an effective introduction. Therefore, discussion of this problem provides a good basis to clarify many of the guidelines introduced in Chapter 2. First, I want to emphasize that the problem as stated without the scaffolding

I will discuss would be unlikely to be effective from an instructional point of view. Specifically, if this problem was given without scaffolding to a typical average academic class without experience solving nonroutine problems most students would either be unproductively frustrated or generate a poor solution that would not support effective processing!

Again, I want to describe in some detail the scaffolding I integrate into this problem to make it at the right level of difficulty. First, the purpose of the scaffolding needs to be clear: The scaffolding should transform a problem that would be unproductively frustrating without the scaffolding into a problem that would be appropriately challenging and provide a good basis for fruitful processing. Notice that the purpose is not to make the solution of the problem straightforward and to insure that all groups determine the best answer – such a purpose would many times be appropriate if our focus was facilitating students mastering a content objective but not if our focus is on improving the student's ability to solve nonroutine problems. In the latter case, we want to immerse the student in an appropriately challenging experience attempting to solve a nonroutine problem and facilitate good processing of their work; thereby making it likely that they will naturally improve in their ability to solve nonroutine problems.

The major components of the scaffolding are:

(1) I use a visual aid (taped rectangular solid) to demonstrate the dimensions involved in the 30″ of tape, particularly making sure they understand the definition of a girth. I would not want a misunderstanding of "girth" to prevent the students from productively engaging with the problem.

(2) The students work in cooperative groups for this task. A cooperative group model is especially appropriate for an introductory problem, considering that students generally have not had much or any experience with solving problems involving the three steps. For this problem, to facilitate interdependence I randomly assign the following roles: (a) calculator, only one student in the group has a calculator and can use it to generate data, (b) recorder, only one student in the group has a writing utensil and can record data and (c) facilitator, one student keeps the group on task and focused. Of course, any student in the group can suggest/tell the calculator what to calculate and the recorder what to write. These roles increase the likelihood that the students will work together on the problem.

(3) I demonstrate a procedure and format to generate data: they pick a length, subtract from 30 (equals a girth) and divide by 2 (equals half of a girth, which is the sum of a width and a height) and then enter

a width and height equal to that sum, and calculate the volume. For example, if you pick a length of 12″ then a girth equals 30″ −12″ or 18″. Next, 18″ divided by 2 equals 9″, which is half of a girth, which is the sum of a width and a height. We can then take for the width and height of the box any two numbers (at least one a whole number) whose sum is 9″ (e.g., take the width equal to 6″ and the height equal to 3″). Notice, in this case, the length equals 12″, width equals 6″, height equals 3″ for a volume of 12″ × 6″ × 3″ or 216 cubic inches, using exactly 30″ of tape (i.e., length = 12″, girth = 2 × (6″ × 3″) = 18″, and 12″ + 18″ = 30″).

(4) In demonstrating the procedure, I use the above example (12″ × 6″ × 3″ = 216 cubic inches) and 6″ × 10″ × 2″ (equal to 120 cubic inches) with the following emphases: (a) helping students notice the large difference in volume between the two solids; that is, 30″ of tape does not determine the volume of the solid; (b) pointing out in the second example that the length does not need to be the longest dimension; (c) letting the students know that one purpose for showing the two examples is to insure that they can generate data when they work on the problem, but not to give them clues about the best solutions. Specifically, I make it clear that I do not want them at this point to identify better choices for the three dimensions or to start experimenting with additional examples and checking for understanding of the process during the second example by asking students what the next step would be and leaving the two examples on the board so that students can refer to them during their group work.

(5) I give the following hints to students concerning how to work on the problem: (a) I remind students of the essence of the step of trying something that suggests for this problem that they generate a good number of examples looking for patterns. (b) I point out that in my two examples a better approach would be to start with a length and then look at a variety of examples of girths (e.g., for the example for 12″ length, in addition to a girth with 6″ and 3″, we could try 5″ and 4″, 7″ and 2″, 8″ and 1″, rather than just one pair of width and height for the length). (c) I let students know that once they generate a good amount of data, they may discover two patterns that will reduce the number of examples they need to try and help them determine the most likely examples to find large volumes. (d) I indicate that each of their three answers should have a volume that is more than 216 cubic inches.

(6) I point out to the students that the requirement of exactly 30″ of tape is to make the problem more challenging by requiring three unique

solutions. For example, if 12″ × 6″ × 3″ = 216 cubic inches was a best solution, 12″ × 6″ × 2.99″ would be a second trivial solution with a minor change in one dimension, using slightly less than 30″ of tape. Also, I tell them that the significance of the second restriction, "have at least two of the three dimensions (length, width, and height) be a whole number of inches," is similar to the information in the Census Taker Problem that "the oldest child likes chocolate ice cream" in that initially the restriction may seem strange (e.g., why two dimensions versus one or three?) and not useful in solving the problem. I indicate that just as the significance of "the oldest child likes chocolate ice cream" only becomes clear when you persist in solving the problem, the significance of the requirement to "have at least two of the three dimensions… be a whole number of inches" might only become clear after persistence.

(7) I remind the students that they are to persist working on the problem until they feel confident that they have the three best solutions (no time limit) and that when they believe they have the three best solutions they are to turn in one paper for each group with the three sets of dimensions and a rationale why they believe they have the best solutions (one to two paragraphs). Although they are given unlimited time, reasonable adjustments are made; for example, if a group seems to be pursuing a truly unproductive approach, I might give them a hint that readjusts the difficulty level. Also, if all but one or perhaps two groups are satisfied with their solution and have completed or are in the process of completing their written justification and the remaining groups are not close to a good solution, I might limit them to a few minutes to identify their best solutions (without a written rationale). In theory you want to give each group enough time to complete their solution; however, you also want to have enough time to adequately process their work.

(8) When students begin to work in groups, I limit my help to two items. First, I check students' computations to insure that they multiply correctly and satisfy the 30″ requirement (e.g., sometimes students take two numbers for the width and height whose sum is the girth versus half of the girth) and have them correct any inaccurate data. Second, I make sure that they are generating a good amount of data, enough to make it feasible for them to discover the two patterns. Some groups tend to discuss what "logically" would be the best solutions, rather than actually generating data that might indicate patterns. Notice that I do not tell them if their solutions are good or the three best solutions. A major requirement in the problem

from a pedagogical point of view is that they need to determine if their answers are the best and if it is the right time to hand in their solution. This approach facilitates a good discussion of the step of persistence.

(9) Occasionally, I will give a group more direct help, such as indicating some data to generate, when it seems that they are unproductively frustrated or their approach is unlikely to give them the experience that would make the processing fruitful.

The processing usually helps students not only see how trying something results in uncovering patterns that make the problem easier, but also notice how persistence can lead to the best three solutions. For example, usually contrasting the approach of groups that identify three of the five best answers (e.g., 250 cubic inches and the two solutions for 240 cubic inches) with groups that identify the three best solutions by including lengths of 9 and 11 inches allows for a good focus on the third step of persistence. Given the help described, most groups are successful with this problem, meaning not necessarily that all the groups find the best solution, but rather that most of the students are engaged in working on a problem at the appropriate level of difficulty and participate in the processing of the problem. A good indication that their problem solving abilities will naturally improve (e.g., Caine, 1994).

If I assigned this problem later in the year after students had considerable experience solving nonroutine problems, I would reduce the amount of scaffolding. For example, I would not check their calculations for accuracy. Later in the curriculum if they generate examples with volumes inconsistent with the patterns and other data, then that should be a clue to recheck the accuracy of the data! Also, considering that there is one correct answer to this problem and that I give the students a procedure to generate data, my approach may seem to contradict the characteristic of a nonroutine problem, "The problem is open-ended; that is, it allows for various solutions." From the point of view of the pedagogy of the curriculum, "open-ended" concerns the experience of the student more than whether there is only one correct answer to the problem. For example, with the given scaffolding, students' solutions generally fall into three categories, the three best solutions (perhaps 25% of the groups); three of the top five solutions (e.g., 250 cubic inches and the two solutions for 240 cubic inches) and a few solutions with less than three of the top five solutions. This distribution is excellent in that it provides the basis for productive processing and increases the likelihood that students will improve in their ability to solve nonroutine problems. Finally, the characteristic of a nonroutine problem, "Every student is able to 'solve the problem'. Of course, the quality of different student solutions will vary, but students will be able to confront the problem and generate a solution consistent with their ability and efforts" suggests we need to provide the student with the procedure to generate data to insure that the problem is in the zone of optimal proximity for most students.

Additional examples of introductory mathematics nonroutine problems

For the remainder of this chapter, I will discuss additional examples of introductory mathematics nonroutine problems but not provide as much detail as for the Expensive Tape Problem. I will start with four excellent introductory nonroutine mathematics problems that can be completed in one to three class periods (or portions of class periods), not requiring work outside of class. This format allows for more direct supervision by you, a useful characteristic for introductory problems. I will then discuss five additional introductory nonroutine mathematics problems that require some work outside of class but serve well as introductory problems.

Problem 1. "Marbles": In "Marbles" the problem is for the students to predict the percentage of four types of marbles (different colors) in a nontransparent bag. Typically, I put somewhere between 100 and 200 marbles in a bag with a distribution of the four colors between 15 and 35% for each color. Students work in groups and are allowed to draw ten marbles at a time, record the information and return the marbles to the bag. A portion of the assessment is a combination of the group's accuracy (i.e., group's percentages compared to actual percentages) and efficiency (i.e., number of marbles sampled). Typically, after about four rounds, the percentages begin to converge and by about the eighth round the percentages vary little and there is a good basis to predict the final percentages. If the students recognize the general patterns of the percentages (e.g., the reds are slowly increasing such as 34% to 35% to 36%, indicating that the actual percentage is higher), groups are usually off by less than 3% per color (12% total). Groups that do not recognize the general patterns are usually within 20% (e.g., their prediction is the actual percentage for the last round). What I like about this problem is that the students try something by generating and recording data and very quickly "see" the problem become clearer.

To insure that the problem is at the appropriate level of difficulty for an introductory nonroutine problem, I do one demonstration with the class as a whole. I explain what is in the bag, and that I am going to take out ten random marbles at a time, record the data and return the marbles, and repeat that process until the student feels confident to predict well the actual percentage of marbles for each color. I indicate that a good prediction would be off the actual percentages by 12 points or less when you add up the total of how much each color was off. I set up a chart for recording the data which shows for each color the total drawn for that round, the cumulative total for the color and the percent of that color for the total. I make a handout that allows more rounds than is needed, indicating to them that I want them to make the judgment when they have adequate data to make their prediction. For the class demonstration, I give the students a chart and indicate that when they believe they can make a good prediction (e.g., within 12 points) they indicate on a prediction sheet the round number, their final prediction (percentage for each color) and their name. Students can work individually, in pairs or small groups to reach a consensus on when to predict. In practice, I find that for most classes the interaction of pairs or small groups facilitates the type of discussions that provide the foundation for good processing. For the demonstration, we record the data as a class with me demonstrating how to convert the cumulative total into a percentage, a skill they will need for the task after the initial demonstration. If a few groups have not turned in their predictions and the data is obviously adequate you can ask the remaining group(s) to submit their predictions. For the processing, I usually like to add about two more rounds of data to see if the additional rounds actually make a difference in the prediction accuracy. Usually, the difference is not significant. In the processing of the demonstration round, I like to concentrate on two focuses: (a) the fact that trying something (recording data) quickly makes a difficult problem easier, and (b) issues concerning persistence, such as did students predict too quickly, and did some groups wait too long? Discussing the actual data and when it began to converge, and when it stabilized is obviously appropriate, particularly noting if the extra data added after the last prediction was necessary. Typically, one demonstration is adequate. More than one demonstration may be appropriate if the data or if the students' understanding seemed not to be adequate in preparation for the next round.

For the next step, I form students into small groups, giving each the task of predicting the distribution for their bag. I organize this effectively for a class with no other help by having four or more bags, each with a different distribution to reduce the likelihood of students using data from other groups and allowing four groups to sample at a time and letting students (one per group) draw the marbles and record at my table. If another person is available to help, the classroom management can be easier. The processing of the problem consists of two steps: (a) reporting the results, the actual percentages and the accuracy of the predictions and (b) relating their experience to the three steps of a nonroutine problem. For example, discussing the fact that generating data made the problem easier. The first step, reporting the results, allows the students to be able to focus on and appreciate the substance of the processing. Ideally, you want to complete both steps of the processing of the task the day of the task; however, you need to judge whether there is adequate class time and/or if you need time to look at their predictions more closely so you can better organize the processing. While saving the processing to the next day is not ideal theoretically, you need to assess your specific class, considering your comfort level, remaining class time and whether without additional time you can select which results would be most fruitful to discuss. I will mention that "Marbles" is a great problem to help students understand the power of converting to percentage. If one just looked at the patterns of the raw data (without converting), it would be difficult to judge if the data is converging. Therefore, it seems appropriate to include a discussion of the power of percentages in the processing of the problem.

A typical effective outcome might look like 25% of the groups within 12% total for the four colors; 50% of the groups within 20% total (but more than 12%); and 25% of the groups off by more than 20%. This distribution is excellent in that it provides the basis for productive processing and increases the likelihood that students will improve in their ability to solve nonroutine problems.

Problem 2. Maximizing area. The task for this problem is to maximize the area enclosed by a given length of wire enclosure. You can provide a context such as you want to enclose as large an area as possible for a garden with a maximum of 100' of wire (attached to posts). This problem lends itself to more than one version of the problem. As with all the problems, the actual problems you give your class needs to be consistent with the needs of your students. In practice, I find the following sequence of three problems works well, all with 100' of wire, all asking for the shape(s) that has the maximum area with the given perimeter: (a) rectangular shape (maximum: 25' by 25', a square); (b) rectangular shape with one side against a wall (maximum: 50' by 25' by 25', the wall providing the fourth side) and (c) two best shapes for area, without using a wall as a side (maximum: circle, also, the more sides for a regular polygon, the greater the area for a given perimeter). Most students do not realize that for the first problem a square maximizes the area of a rectangle with a given perimeter. Therefore, this problem usually allows students to appreciate the power of trying something in that once they try a few examples it usually becomes clear that the square is the best solution. Generally, little persistence is needed to solve this problem. In contrast, the second problem provides a good basis for processing concerning the step of persistence, in addition to some discussion of trying something. After the first problem, many students assume the best answer for the second problem will be the square (33 1/3'), not realizing that the wall provides "free" extra perimeter; i.e., the square uses only 33 1/3' of wall, while the solution uses 50' of wall, countering the maximizing quality of the square shape. Notice the solution balances these two factors that tend to maximize the area by basically creating two squares next to each other. In the processing of this problem, it can be fruitful to focus on the groups that initially thought that the square was the best answer. Did the group "accept" the square as the best answer without persisting by trying additional examples? Did groups that went beyond the square after trying additional examples, persist until they found the best solution? If yes (or no), what convinced them that their answer was the best solution possible? Typically, this line of processing is very effective in facilitating students understanding concretely the nature of persisting or not persisting.

Also, notice that not trying additional examples beyond the square, provides a good basis for clarifying the first step of problem orientation and the connection with the step of persistence. I like to emphasize that assuming that the solution is the square is not a "mistake," but rather a natural first step, given their experience with the first problem. The "mistake" is in not trying additional examples to verify or reject their conclusion; or, if they did try additional examples, not persisting until they have good evidence that their solution is sound. This is one of the few nonroutine problems that provides a clear example of the significance of the first step. If you do not realize that there is a problem (e.g., that your initial "obvious" solution may not be correct), you are not going to persist. The third example provides for a very good discussion of persistence, given that there are no two best answers. While the circle is the best answer, there is not a concrete best second solution since you can always add an additional side to the regular polygon and get a better second solution. In addition, few if any groups determine that the circle is the best solution, overlooking the circle as an option. This third problem lends itself to a class discussion of what persistence means in practice for a problem for which you will never achieve the perfect solution, given your limited time and resources or the nature of the problem. Of course, there is not one right answer to this question! Therefore, after a variety of student solutions have been discussed, I like to have students discuss the question of persistence in small groups and come to a consensus to what persistence means in this problem and then share their conclusions with the class. One format for this small group task is to have a sample of solutions (including their rationale that their solution demonstrates persistence) from the groups on the board and have the groups rank order the solutions on the criteria of persistence, pick the top three on the issue of persistence, or pick one of a few ratings (e.g., 1 to 5 scale) for each, as well as providing a rationale for their ratings. Depending on your teaching style and the characteristics of your students, you may prefer instead to share your assessment of this step by discussing a few examples selected from the class's solutions of good progress on understanding this step. In facilitating this processing, if appropriate,

I like to remind the students of the abstract characteristic of this step of persistence, that you feel convinced that you understand the problem as well as is possible for you. For example, if you understand that the circle is the best solution, that regular polygons maximize the area for a given number of sides and that the more sides the polygon has, the greater the area, then you have an excellent understanding of the solution and have demonstrated an understanding of persistence. Further, I would argue that even if you did not understand that the circle is the best solution but did understand that regular polygons maximize the area for a given number of sides, and that the more sides the polygon has, the greater the area, you have demonstrated a good understanding of persistence. You demonstrated a clear qualitatively better solution than when you started the problem, even though you did not consider the possibility of an enclosure that was not a polygon. In the processing I like to emphasize that the concept of persistence is relative in that the affective component is a change from feeling that you do not have an acceptable solution (and need to keep engaged with the problem) to feeling that you now have a qualitatively better solution or understanding of the problem and it is the time to complete your solution.

How you organize work on this problem depends on your specific class. For example, some factors you might consider:

(1) Should students work on these problems individually or in small groups? Although I generally recommend small group work for the introductory problems, this problem can be an appropriate introduction to individual work if they have had some experience working in groups with other introductory problems.

(2) Should the problems be worked on in the same time period or over a few class periods?

(3) For the first two problems, is one specific problem enough? For example, after wire of 100', would an additional example be productive (e.g., 200' or one that requires a fractional side, such as 150')?

(4) What support do you give the students for the third problem? For example, do you demonstrate how to determine the area of a regular polygon, or do you demonstrate how they can use graph paper to approximate the area, perhaps using string for the given perimeter?

(5) Is it appropriate to give one or more of the problems as an individual assignment outside of class? For example, after completing problem one in class, perhaps giving two problems as homework assignments, one just like the first problem (e.g., the same except 200' perimeter) and problem number two, perhaps giving the students the option of trying the problems alone or in pairs.

(6) Is the third problem too difficult for this class as an introductory problem? Should you omit the problem, or facilitate a class discussion and experiment that guides the students through the process with your support? In answering these questions, the key principle is to ask yourself if the sequence and format you use is likely to engage students in working on problems at the appropriate level of difficulty. If you can answer "yes" then, with proper processing of their work, the students are likely to improve in their ability to solve nonroutine problems.

Finally, you need to allow yourself to be open to the need to adjust your original sequence based on what you observe as the student's progress through the sequence.

Problem 3. "Estimating the number of beans." In this problem, students are given the task of estimating the number of beans (e.g., pinto beans) in a large jar. I make sure the beans are not level, but rather on an incline to make the problem more interesting. The only restriction on their methods is that they are not allowed to touch the jar. For example, they are allowed to use a ruler as long as they don't touch the jar (e.g., hold the ruler about an inch from the jar to measure); however, they cannot shake the jar to make the beans level. This is an excellent nonroutine problem to introduce students to the step of persistence in solving a nonroutine problem. For example, without experience solving nonroutine problems, most students determine their estimation in one of the two ways: (a) they determine one method to estimate and use that method to determine their final answer or (b) they use more than one method and average their answers to obtain their final answer, even when the two estimates are not similar. The key characteristic of the step of persistence is that you continue to engage with the problem until you feel confident you have an excellent answer or the best answer that you are capable of determining. In both cases, there is a significant difference in the quality of your understanding of the problem. Therefore, accepting your first attempt as the best possible estimate is inconsistent with the step of persistence. Averaging your answers when the estimates are not similar is even more inconsistent with the step of persistence. From the perspective of this step, the natural questions to ask when your estimates are significantly different are "Why are my estimates different? Can I redo the methods or try additional methods until I determine an estimate for which I feel more confident?" I remember a group that was working on the problem of estimating the number of leaves on a very large maple tree and for their final answer they averaged two estimates that had a difference of more than 200,000 leaves and did not see a need to reexamine their work!

As implied above, a major purpose of this nonroutine problem is to strengthen students' understanding of the step of persistence. There are a variety of methods to organize the pedagogy for this problem. There are two general options for the directions for the task that I have found useful. First, you can require each group to attempt at least two methods to estimate before they make their final estimation. Typically, with that approach many groups will either use one of the estimates as their final estimation or average the two estimates, providing an excellent basis for processing the task and contrasting approaches' consistency with the step of persistence. Second, you can give the directions with no indication or requirement concerning how many estimations to make, just emphasizing a final estimation that the student group feels confident is excellent. Typically, in this approach, many groups try just one method, providing a focus for the processing. If a few groups (or even one group) use an approach that exemplifies good persistence, their approach can be contrasted to other approaches.

What these two approaches have in common is that the students are given a nonroutine problem to work on with enough scaffolding to help insure that the problem is at the right level of difficulty, but not enough to insure that all (or almost all) groups will generate an excellent estimation. The general format that is most productive in improving students' understanding of the process of solving nonroutine problems is one in which: (1) most students are actively engaged at working on a problem at the appropriate level of difficulty, (2) approximately half the groups (or students) generate a reasonably good solution, a quarter generate an excellent solution and a quarter generate a less adequate solution and (3) the processing of the task focuses on helping the students connect their experience solving the problem with other student's approaches and your comments concerning their work. Our approach to grading can provide a fruitful line for the processing. For example, if you have two groups whose estimate is close to the actual number of beans, but one group's estimate was based on good persistence and the other group was just "lucky" (e.g., used just one method without a rationale to justify confidence in the estimation), I assign the former group a grade in the A range, and the latter in the B range. Of course, this usually generates a very in-depth discussion of why this approach to grading is valid!

Another option, especially if you believe your students need more concrete scaffolding to make the problem at the right level of difficulty, is to work with each group individually by requiring them to submit their work to you in two steps, first, having students submit progress reports concerning what methods they intend to use and why they believe they are accurate and, second, students submit their final estimate and explain why they believe it is adequate. For both steps, you have the option to require them to resubmit based on additional scaffolding you provide. Of course, you can format those two steps as a whole class discussion rather than working with individual groups. In addition to selecting an approach that you believe will make the problem at the right level of difficulty for the students, you need to select an approach that you feel comfortable implementing. At the same time, it is important that you move in the direction of less scaffolding for the students over time, meaning that at times you need to be willing to experiment with formats that are at the appropriate level of challenge for your context.

The "Estimating the number of beans" problem can be a good problem to facilitate students making the transition to problems that require them to work outside of class. For example, the students can have one class to talk about their strategy, inspect (or measure) the jar, and plan what they can do outside of class to improve the quality of their solutions, including allowing them to bring any materials to the next class in which they are required to make their final estimation. For example, some groups I have worked with bring a similar jar and/or beans to the class, in my opinion, an excellent strategy that is not counter to the rules and which can generate a good discussion of how we can solve real problems with clever strategies.

Finally, this problem lends itself to similar problems that can allow the students to deepen their understanding of the process of persistence. For example, you can vary the type of container (e.g., a more irregular shape) or the types of objects in the container. In practice, I have found that usually one or two problems of this nature is adequate to insure that most students improve in their understanding of persistence.

Problem 4. "Calculating the area of an irregular shape." This nonroutine problem requires students to estimate the area of an irregular closed curve drawn on graph paper. In practice, this is a good problem for introducing students to the three steps of solving a nonroutine problem for two reasons. First, this problem requires very little content prerequisite skills and is a problem that can fairly easily engage most students. Unlike many other nonroutine problems there is seldom difficulty in getting started by "trying something," although the quality of students' efforts can vary from poor to outstanding. Second, the differences in the quality of their solutions form a productive basis for better understanding the step of persistence. For example, a common solution involves counting all the complete squares (1 unit by 1 unit) and then estimating the remaining area by rounding up or down for each square partially filled (e.g., 1 square unit if mostly full) or another straightforward counting strategy. Students that demonstrate a better understanding of the step of persistence, develop more sophisticated strategies, such as dividing each partially filled square unit into ninths and then estimating, or cutting all the partial square units (filled part) and putting them together on the graph paper to fill units. One group of calculus students cut out the irregular closed curve and the entire rectangle around the curve, and then using very sensitive science scales, set up the following proportion: the area of closed curve divided by the area of the known rectangle equals weight of closed curve divided by the weight of the known rectangle, and then solving for the unknown area.

In introducing the problem, you can explain that their assignment is to calculate the area on the handout as best they can. You might use an analogy such as the following: "Your company is competing for a job with several other companies. The company that calculates the area closest to the actual area will receive the job."

There are a number of formats for this problem that facilitate students better understanding the process of solving nonroutine problems. One format that has worked well for me includes the following components:

(1) You can assign this problem as an individual problem. As stated previously, unlike many nonroutine problems, most students can engage with this problem without the support of a small group and with little or no scaffolding; therefore, this problem provides an excellent opportunity for the student to experience working on a nonroutine problem at an introductory level individually.

(2) I suggest making a central focus of the processing discriminating between solutions that demonstrate a good understanding of the step of persistence and solutions that do not, perhaps adding some suggestions that seem within their zone of potential understanding. For example, one student approached determining the area of the irregular closed curve by calculating the average width times the length and ending up with an answer of 286.2 square units. Then the student calculated the area similarly, but by determining the area of the outside first and then subtracting from the total area. Using this method, the student obtained an area of 313.6 square units. The student's final solution was the average of the two values. The student obviously did not evaluate the adequacy of the answers. If the techniques were valid and carried out correctly, how could the student get two answers so different? What was needed was persistence to discover the minor mistake that threw the one calculation off.

(3) You can assign a second similar problem with the understanding that students would need to demonstrate an improved understanding of the step of persistence to receive a grade of B or better. This second problem provides the opportunity for the student to "practice" the step of persistence with scaffolding (e.g., completing a similar problem

with processing). A variation of this format is to require at least two (or three) strategies to be attempted before coming to a final solution. An advantage of this variation is that it increases (actually guarantees) the likelihood that the student will try more than one strategy, hopefully thereby improving their understanding of persistence; a disadvantage is that it tends to remove the natural understanding that develops through seeing other approaches and comparing to yours. When I have tried this variation, I have noticed that the second strategy of students tends to be included to satisfy the requirement rather than a strategy developing out of an understanding of the need to persist. Or some students will average two solutions with no reasonable rationale, again a step inconsistent with the step of persistence. However, these errors in reasoning due to the requirement of more than one strategy can become the basis during the processing of understanding the step of persistence more clearly.

One implication of the above comments is that even when our requirements or directions result in solutions inconsistent with our intentions, good processing can help the students see more clearly how to implement the steps in future problems.

Two important concepts to emphasize in the processing are the variety of techniques that the students discovered and the role of persistence in this problem. You can focus on the variety of techniques by sharing as many different good methods as possible from the students' reports, trying to involve methods from as many of the students as you can. It can be valuable in the processing to go over examples of students both persisting successfully and not persisting. In discussing examples of not persisting, you might emphasize that the student discovered an excellent method and how, with persistence, that method could have resulted in an excellent answer.

Problem 5. "The most pleasing rectangle": The task for this problem is to construct the rectangle that will be judged most pleasing by other people. What I particularly like about this problem is that initially it seems to the students not to have a solution and to have nothing to do with mathematical problem solving, but after the problem is completed and processed it becomes clear that the three steps of solving nonroutine problems result in fairly consistent good solutions. Basically, what happens is that the solutions judged to be most pleasing tend to have a ratio of adjacent sides close to the Golden Mean.

To insure results consistent with the Golden Mean I place the following restrictions on their rectangles: (a) the rectangle must be made from a piece of black (or any consistent color) construction paper with no designs on the paper and (b) the rectangle must have one side equal to six inches (or if you prefer a metric measurement). These restrictions help insure that factors other than the shape of the rectangle do not effect whether the rectangle is judged pleasing or not. The restriction of one side being six inches helps make the number of possibilities manageable. Unless the students are experienced solving nonroutine problems, I suggest that they start the problem by remembering the second step of solving a nonroutine problem, trying something. For example, I suggest that they cut out a variety of rectangles consistent with the restrictions, show them to a variety of people, and see if any patterns emerge concerning which rectangles are judged pleasing. I also remind them of the third step of persistence, noting that the first set of rectangles may suggest some patterns but that an excellent solution will require the student to refine the experiment until the results are consistent and convincing. I usually add that "Believe it or not, there is a 'correct' answer to this problem and I will predict on a piece of paper in a sealed envelope which rectangle will win."

Although this problem can certainly be done in cooperative groups, I prefer assigning the problem individually because I find that most students can have success with this problem if they follow the advice given in the directions, thereby individually appreciating the power of the three steps in solving nonroutine problems. Alternatively, the problem can be assigned as a small group task. This approach reduces the number of rectangles that need to be evaluated, as well as requiring students to agree on a strategy as well as when they have persisted adequately. The students are required to turn in one rectangle to be judged and written documentation of the process and data they used to determine their rectangle. When the students have constructed their rectangles, I tape them to a suitable wall or blackboard and have each student in another class pick the most pleasing rectangles (usually I have them pick approximately 15% of the rectangles). Sometimes I include art teachers, artists or art students as the judges. We then tabulate the data to determine the most pleasing rectangles. If there is time, having the students solve the problem of how to determine which rectangle is most pleasing gives the students additional excellent practice in problem solving.

Invariably the rectangles judged to be most pleasing have a ratio of the long side to the short side reasonably close to the Golden Mean (the short side equal to six inches usually is judged more pleasing than the long side six inches). Since the Golden Mean is approximately 1.618, this means that if the short side is 6", the long side would be approximately 9.7", and if the long side is 6", the short side would be approximately 3.7". Generally, the students that persist in their experiments are more likely to determine values close to the Golden Mean. Therefore, in the processing of this task, I like to emphasize how the steps of trying something and persisting allow the problem solver to determine a good solution to what seems like an impossible problem to most students. After selection of the most pleasing rectangles and discussion of how the steps of trying something and persisting are valuable in solving this problem, I have them determine the ratio of the long side to the short side of the rectangles judged most pleasing. Then I reveal my prediction that the ratios would be close to the Golden Mean and discuss some of the mathematics involving the concept of the Golden Mean.

In assessing this problem, I usually assign 50% of the grade based on how their rectangle is judged and 50% on the quality of their effort as documented in their written report. If a student demonstrated a good effort in constructing sound experiments and persisting, I assign a grade of at least 80% even if the rectangle is not judged as particularly pleasing.

The specifics of the directions need to be adjusted to be consistent with the primary purpose of this introductory problem: To introduce the student to the three's steps of a nonroutine problem in a way that facilitates them being successful with nonroutine problems with less scaffolding. Remember, for an introductory problem most students are likely to need help seeing what is appropriate for the steps of trying something and persisting. For example, you might give them specific directions such as: for the first step, create at least five different rectangles, with a difference of at least one inch (or 3 centimeters) for the second dimension of the rectangle, and gather data from at least 10 people. You might indicate what type of data to gather (e.g., two most pleasing rectangles; rank order the rectangles; etc.). You might process this first round of data as a class by having each student identify their most pleasing rectangle and looking for patterns. For example, rectangles with a second dimension between nine and ten should be judged most pleasing more than other dimensions. This type of outcome provides a good basis to discuss the importance of persistence. Specifically, you might indicate that the results imply that the best answer is between nine and ten, and ask how can we persist so that we feel more confident in our results? Depending on your assessment of your class, you might either come to a class decision; e.g., compare additional rectangles with given suggestions or directions such as compare using 1/8" intervals (9 1/8; 9 2/8; 9 3/8; 9 4/8; etc.) or let them decide individually how to persist, perhaps allowing them to work on this step in small groups.

Problem 6. "Maximizing volume": For this nonroutine problem, the students are given this problem:

> Your group is a research team for a company. You are given the following problem: The company sends out many containers and the company wants to save money by fitting more material in each container.

The containers are made out of a sheet of material the size of a piece of construction paper, and do not have a lid (different material is used to cover the top of the container). You can design any shape as long as the top open section (where the lid would be) is flat. The company wants your three best designs for a container to choose from; each design should be a different shape.

Let students know that they can make use of masking tape to construct the container as long as the tape is not used as additional surface area (or is a minor addition to surface area). In addition to scissors, construction paper, and tape, you may want to also make some material available for them to test their volumes, such as a bucket of sand and a measuring cup.

This is an excellent introductory problem that students can work on in groups at least partially outside of class, and, as in the "area of an irregular figure," a problem that most students do not have difficulty starting; therefore, providing a transition from introductory nonroutine problems that are completed in class to ones that require work outside of class. Before they start, you do need to make sure that they understand the restrictions of the problem and also the variety allowed, they are certainly not restricted to rectangular solids. It helps to have them generate one or two practice "solutions" in class and compare their results to make sure they understand the restrictions of the problem and get a sense of the range of volumes (e.g., how many cups of sand can the container hold?). At least two practice models help insure that the practice ones provide a variety of potential shapes, versus all rectangular solids.

What I especially like about this problem is that it provides an excellent opportunity for students to see directly what it means to persist in solving a nonroutine problem through additional exploration after their initial solutions. For example, three common choices for containers are a rectangular solid, a cone and a cylinder. After the processing, you can give them a concrete example of persisting by selecting one of these shapes and asking how we can be confident that we have picked the rectangular solid (or cone, or cylinder) that holds the most volume. You can adjust the instructions depending on your assessment of their understanding at this point. For example, as a class you can systematically try rectangular solids with square bases (maximizes volume for a rectangular solid with a given perimeter for the base) with sides of 1", 2", …, 8" by having each group construct one. Experimentation will indicate that the best volume is when the side of the square base is between 5 and 6" for an 8.5 × 11" piece of construction paper. Depending on the understanding of your students, you can move in a variety of directions: (a) exploring further the rectangular solid, perhaps asking what would be reasonably appropriate in terms of persistence; e.g., dividing into fourths, tenths?; (b) exploring in small groups or individually cones and cylinders; (c) exploring how to persist in answering the general question of the best three shapes; (d) exploring how you can approximate the best theoretical shape, half of a sphere, with the materials you have. For example, one exceptional calculus student cut out 20 long strips in the shape of a right triangle with a small base and put them together to approximate half a sphere, obtaining a volume significantly larger than the other shapes submitted for that class.

In these follow-up explorations, some more subtle concepts concerning the step of persistence can be considered such as, "How do you discriminate between reasonable persistence and spending an inappropriate amount of time or effort on a problem?" For example, in exploring which base is best for the rectangular solid, there is not a need initially to take too much time in constructing the different solids, the models just need to be accurate enough to allow you to see the pattern concerning when the volume peaks. Also, for example, in the follow-up problem to determine the maximum value for the volume of a rectangular solid, the processing can focus on the issue of persistence by comparing three solutions, the volume when the length of the base for the maximum volume is determined to the nearest (a) whole number, (b) quarter of an inch or (c) tenth of an inch. I would suggest that (a) nearest whole number, does not indicate excellent persistence, that (b) nearest quarter of an inch, does indicate excellent persistence and that (c) nearest tenth of an inch is an example of persisting more than what is needed at this level. Finally, the question of how many shapes to explore before you feel confident that you achieved an outstanding solution with your given resources and time provides an excellent forum for gaining a practical sense of the step of persistence. When I explore these questions with students, I believe it is important to frame these questions as open-ended questions; that is, there is not a specific "correct" answer, other than the general principle that you should feel that you have generated the best solution you are capable of constructing with your limited time and resources.

Problem 7. "Graphing": "Graphing" requires students to determine the graph (nonlinear) of the solution set of an equation, given the equation. Each graph requires students to try some values, look for patterns, and persist until the entire graph is clear. From the point of view of a sequence of nonroutine problems, "Graphing" is a powerful demonstration for students of using the three steps in a fairly straightforward manner to solve a difficult problem, an experience most of them have not had. They are given equations to graph that probably seem very difficult or impossible to them; however, by applying the three steps, the graphs "miraculously" gradually become clear.

Before attempting this problem, the students need to understand the connection between an equation and the graph of the points that satisfy the equation and how to plot points on a x-y coordinate system. Notice that it is not necessary for students to know how to graph or determine the equation of a straight line from two points. In fact, I would argue that this problem is an excellent introduction to graphing that would increase the likelihood the student would understand the concept of graphing both linear and nonlinear equations, especially the concept that a graph of an equation represents the solution set for that equation.

This problem when used as an introductory nonroutine problem requires a good amount of teacher support to make the problem at the appropriate level of difficulty. Most classes need an introduction to the process of using the three steps to graph before starting the problem. I demonstrate the following process to students: try something by substituting some easy values (e.g., $x = 0, 1, 2, 3$ and 4), look for patterns indicating sections of the graph that seem clear and sections that do not seem clear, try additional values to check the clear sections and to investigate the unclear sections and repeat (or persist) until you feel confident that the graph is clear. I demonstrate the following examples to clarify the process and to prepare students for the type of problems they will attempt to graph $y = x^2$; $y = |x + 2|$; $x = y^2$; and $4x^2 + y^2 = 36$. For example, to demonstrate $y = x^2$: Start by substituting $x = 0, 1, 2, 3$ and 4 and solving for y, then plotting those points. Ask the students which sections of the graph seem clear (e.g., $x > 0$) and which sections need more investigation (e.g., $x < 0$). You can ask the students to predict the part of the graph for $x < 0$ (usually students will suggest about three different predictions that visually make sense). Try additional values such as $x = 2.5$ and 10 (verifying the clear parts of the graph) and $x = -1, -2, -4$ and -10 until the entire graph is clear. In reviewing the process, point out that the number of points needed to be clear about the graph will vary with the student and the equation. You can demonstrate the concept that the number of points needed will vary by having students predict the entire graph when they feel confident and discussing the accuracy and rationale for their predictions. For example, some students may not need to substitute any values of x less than 0 because they realize that the graph of $y = x^2$ has to be symmetric to the y-axis; others may require quite a few values before they feel confident that their graph is correct. What needs to be emphasized is that you substitute values until you are confident that the graph is correct, that is the essence of the third step of persistence for this problem and that when you feel confident is a personal experience that only the problem solver can determine. The second demonstration problem helps insure that students understand the steps for substituting into expressions involving absolute value, the third introduces the strategy of substituting for y instead of x and the fourth demonstration emphasizes how to determine points for an equation for which substitution for either x or y does not always lead to a simple calculation of the second variable, including strategies concerning how to "spot" easy values to substitute (e.g., x = 0 or y = 0), the strategy of finding four related points at a time (e.g., (x,y), (−x,y), (x,−y) and (−x,−y) and how to determine or estimate values that are not rational (e.g., square roots). In summary, students generally need both a sense of how to apply the steps in this problem and an understanding of some of the specific methods for trying something for certain types of equations. However, in order to keep the problem at the appropriate level of difficulty, you should avoid discussing general features of types of graphs (e.g., the relationship between the graph of f(x) and $|f(x)|$, the shape of even and odd functions, the relationship between the graph of f(x) and its inverse). Of course, a discussion of these general features of types of graphs would be appropriate as part of the processing of their work.

Typically when I assign "Graphing" I have students work in groups, each group is assigned four problems from the 16 given below, one each from each of the following groups of problems: problems 1 to 4, problems 5 to 8, problems 9 to 12 and problems 13 to 16. Graph the solution set of the following equations: (1) $y = x^2 - 2$, (2) $y = x^3$, (3) $y = x^4$, (4) $y = 2x^2$, (5) $x = y^2 + 3$, (6) $x = y^3$, (7) $y = |2x - 8|$, (8) $y = |3x + 9|$, (9) $y = x^2 + 4x$, (10) $y = x^4 - 4x$, (11) $y = x^2 - 6x$, (12) $y = x^3 - 12x$, (13) $x^2 + y^2 = 36$, (14) $x^2 - y^2 = 36$, (15) $y = |x^2 - 9|$ and (16) $y = |x^3 - 8|$. One format that has worked well for me is to have the students work in groups of three in class (the day after the demonstrations to allow enough time for working on the graphs), beginning with an approximately ten-minute period in which each person in the group works individually on one of the first three graphs. After the initial period, each student shares their progress on their graph and students discuss each graph until all three agree on an answer. For this introductory problem, I include the additional scaffolding of the teacher checking the correctness of the points they determine through substitution, insuring that the points they determine are actually points on the graph. I do not indicate to them whether their completed graph is correct or not. This scaffolding helps insure that the focus is on the steps (versus their ability to solve for points), especially the process of persisting until they are confident they have a correct graph. With this format most groups correctly graph two to four of the graphs, only a few groups correctly graphing the fourth example, an excellent distribution for a focus on the process of problem solving.

Other formats that may work better for your context include: (1) complete one graph at a time over four days, either individually or in groups; (2) as part of the process, assign some graphs as homework after completing the first group of graphs and (3) omit graphs from the fourth type, except perhaps for extra credit. This last option is particularly appropriate if you believe the difficulty of the fourth group will distract many of your students from the main objective of the problem; that is, to experience the power of trying something and persisting.

This problem is one of a few nonroutine problems that I have used with first and second year secondary students that requires such an investment of time to insure that relevant content objectives are understood before work can begin on the problem. My experience indicates that introductory nonroutine problems requiring minor or no mathematics content prerequisite skills typically are more effective in allowing a focus on the process of solving nonroutine problems. However, in my experience using "Graphing" with my students, I have found that the above demonstration generally adequately prepares the students for tackling the problem and the "content focus" of the problem does not interfere with their ability to engage in the process of solving this nonroutine problem. In addition, I have found the problem to be an excellent introduction not only to the three steps in general, but also to how the three steps can be used to solve seemingly difficult mathematics problems related to the secondary mathematics curriculum.

I will also note that this is one nonroutine problem that easily lends itself to an assessment of their understanding of their work on the problems. For example, after processing of the projects, I have given the students the following four equations to graph to assess their understanding of their work: $y = x^2 + 1$; $x = |2y - 1|$; $y = x^2 - 6x$; and $x^2 - y^2 = 9$. It should be clear that this is only an assessment of their understanding of the specific problems covered in the assignment and processing, basically indicating whether students understood the specific problems assigned (i.e., a measure of individual accountability for their work and attentiveness during the processing) rather than a measure of their ability to solve a new nonroutine problem.

This problem is meant to be completed without the use of graphing calculators. Problems later in the curriculum assume the use of a graphing calculator; however, the use of a graphing calculator for this problem could interfere with the major objective of the curriculum to improve the student's ability to solve nonroutine problems (see comments below). Therefore, you need to take measures to insure that graphing calculators are not used. Some options include: (a) if it is unlikely that students would have graphing calculators or realize that graphing calculators could be used for this problem, then give the directions without mentioning graphing calculators and emphasize the need to show the sequence of points used to determine the graph, (b) have the students complete the graphs in class, one a day without graphing calculators or (c) directly mention that graphing calculators are not to be used (and the rationale why).

"Graphing" illustrates some of the key implications concerning a curriculum of nonroutine problems. For example, if we assume that students have learned how to graph linear equations but have had little or no experience graphing nonlinear functions, and no experience with graphing calculators, we can contrast two different approaches. First, if our primary objective is to improve the students' ability to solve nonroutine problems, then the above problem without a graphing calculator would be an excellent vehicle to improve those skills and the technology of graphing calculators would not be indicated. In contrast, if our primary objective is to have the students study parabolas, particularly the effect of a, h and k in the equation $y = a(x - h)^2 + k$, the graphing calculator would be an excellent tool not only to discover the patterns, but also to emphasize certain problem solving skills such as finding patterns. The problem of determining the effect of a, h and k in the equation $y = a(x - h)^2 + k$ could be formulated as a nonroutine problem; however, if your primary objective is the content objectives and you had limited time, the format of a nonroutine problem could be inappropriate.

Problem 8. "Grading": The nonroutine problem "Grading" involves analyzing data for patterns, specifically the raw scores from a mathematics test. From the point of view of a curriculum of nonroutine problems, what is more important than the understanding of, and the ability to apply, the key terms and concepts of data analysis and statistics is the understanding that given certain conditions fairly easily recognizable patterns emerge in data that can guide us in analyzing and making sense of the data. In other words, I believe that what is essential in the study of the "mathematics" of data analysis is the understanding that in the process of solving many problems it is useful to generate data (e.g., try something), organize it and look for patterns that help you "construct" the mathematics necessary to answer your question (e.g., persist). The "Grading" problem is an excellent introductory nonroutine problem for two reasons. First, the students are generally successful solving an initially seemingly difficult problem by using the three steps and, based on their experience and the processing of the problem, improve their understanding of the three steps. Second, the processing of their work allows you to contrast their effective solutions that required no substantial mathematics content with the "mathematical" solution involving the complicated formula for calculating a standard deviation for a normal distribution (see below), hopefully beginning the process of facilitating the students understanding that not only can they "do mathematics," but also for many problems they do not need to remember or understand complicated mathematical formulas or content.

Before assigning the nonroutine problem of assigning specific grades, I like to facilitate the students becoming aware of the fact that many times in real data there will be fairly clear patterns in contrast to randomly generated data. For example, in "Grading" before assigning the actual grading task, I first ask students to compare the two sets of data below:

Set A	Set B
16	16
21	19
28	21
29	25
32	28
33	30
40	34
40	36
43	39
43	40
44	42
44	46
44	49
49	52
53	56
53	57
56	60
63	63

Initially students are not told that set A is a set of raw scores (total points deducted) on a midterm in Algebra 1, and set B is a fairly evenly distributed set of numbers with the same range of values. Data such as set A tends to be consistent with a normal distribution, demonstrating patterns different from evenly distributed numbers. For example, in set A notice that there are five clusters of scores (16–21, 28–33, 40–44, 49–56 and 63) with the middle clusters tending to be larger and, also, there is some symmetry in the data.

Without scaffolding, most students have difficulty noticing the differences in these two types of data; therefore, I give them a task such as:

Part I. Below is listed four differences in these two sets of numbers. Rank order the four from most significant to least significant. Most significant would best capture the essence of the differences. Give a rationale for your rankings.

    (a) There are more numbers in the 40s in the column on the left.

    (b) The numbers are not as evenly distributed in the column on the left.

    (c) There are repeats in the column on the left.

    (d) The largest gaps between numbers occur in the column on the left.

Part II. All four of the differences above can be inferred from the essential difference in the two columns. Try to capture the essence of the difference in the column on the left compared to the column on the right in one to three sentences describing the properties of the left hand column. The four patterns described above should follow from your answer.

Hint: This question does not require you to use or know any mathematical formulas. Rather, what is needed is to look at the general characteristics of both lists and note how they are different. It may help to look first at each list and notice its general characteristics, then compare the two lists.

I give Part I first and discuss their answers with them before assigning Part II. I will note that the first few times I gave this task I asked students to compare the two sets without Part I of the instructions and found that students' responses for the most part were inadequate to form the basis for productive processing. Even with the addition of Part I and some clarification of the written instructions (e.g., discussion of the meaning of the "essence of the difference"), only a few groups defined differences that formed the basis for good processing. In processing this initial task, I like to emphasize the connection of this task to the three steps of a nonroutine problem, including the observation that many times if we "try something" by collecting data and allowing the data to speak to us (reveal patterns), we can become aware of patterns that help us solve or address a problem. For example, I attempt to insure that they appreciate the characteristic of clustering in set A and the fact that although initially they might not have noticed the clustering, after processing or sooner, the clusters become obvious.

I inform the students that set A is actually raw scores from a midterm for an Algebra 1 class, and that whenever I give students a test with a significant number of items that results in a range of raw scores (e.g., a midyear exam), I grade the tests by listing the raw scores in order, looking for the clusters, tentatively assigning grades consistent with the clusters and then comparing some of the key grades with the actual corresponding test to see if the grades seem valid with my sense of the actual quality of the test (adjusting the grades if appropriate) and that I believe this method for grading exams is both fair and efficient. I note that basing grades for a test of this nature entirely on percentage of problems correct is not a good method, discussing the inconsistency in the fact that 80% correct on a difficult, average and easy test means three different things, and that it would be very difficult to construct a major test for which 80% of the questions answered correctly would mean exactly what you want it to mean (e.g., a grade of "B–"). In addition, you could discuss the process involved in developing major tests such as the SAT or state tests for which the actual scores are not determined prior to administering the test, but only after looking at patterns in the raw scores (e.g., number of questions answered correctly). I inform the students that the nonroutine problem they will attempt is to determine what grades I assigned for the test for set A, and give them the following general directions:

> As mentioned above, this set of numbers is the raw scores on a test given to a math class of 18 students (a midyear exam). The numbers represent the number of problems missed by the students on the test. The grade for 16 missed was 96; the grade for 63 missed was 50. Your task as a group is to determine the grades for the remaining raw scores. You are allowed to ask for the actual grade for two additional raw scores. You should ask for one score at a time. You are not to share scores I give you with other groups or get additional scores from other groups.
>
> Suggestions: (a) The scores do not fit a linear pattern; that is, there is not a proportional increase in score for equal amounts of increase in raw score; e.g., if a change in raw score from 50 to 60 resulted in an increase of grade of eight, it would not necessarily be true that a change in raw score from 60 to 70 would also result in an increase of grade of eight. In fact, looking at differences in scores only will not give you a good result. (b) You do not need formulas to answer this well (in fact, formulas will probably hinder you), rather look at patterns in the clusters of scores and determine a tentative assignment of grades based on your analysis, then pick one score to ask for the correct value, trying to pick a key score, such as one that you are least confident in your estimate or one that seems a key value to know. Based on the new information, redo your grades and again pick a second key number to ask for the grade. After you have your second score from me, make your final guess of grades.

In the discussion of the directions I note, usually based on a student asking the number of items on the test, that the number of test items is not relevant by noting that if I added 1,000 questions to the test that every student could answer correctly the results should statistically be the same.

Students fairly quickly see the need and effectiveness of organizing data and looking for patterns. In addition, they are usually very successful in grading the tests consistent with my grades. For example, using the data in set A, an average Algebra 1 class on the individual test ranged from a total difference from my scores of 4 to 18, with a median difference of eight; an honors Geometry class on the individual test ranged from a total difference from my scores of 1 to 14, with a median difference of four.

Of course, this activity is most meaningful when you use actual scores from one of your classes. If this type of data is not easily available, here are two sets of raw (first set, points deducted; second set, points correct) and assigned grades (from 50 to 96) from my classes: (1) 16 (96), 21 (92), 28 (87), 29 (86), 32 (83), 33 (82), 40 (75), 40, 43 (71), 43, 44 (70), 44, 44, 49 (65), 53 (59), 53, 56 (56), 63 (50); and (2) 75 (96), 73 (94), 68 (89), 68, 67 (88), 67, 66 (87), 64 (84), 57 (73), 55 (71), 54 (70), 54, 43 (62), 43, 43, 42 (61), 41 (60), 30 (50). Of course, I point out to students that the patterns in the data are mathematical in nature and neutral, while the assigned grades, though based on these patterns, are subjective in the sense that I determine the high and low grades; therefore, when I give the problem of determining my grades it is necessary to give a few key values. The fact that my assigned grades are partially subjective, though basically consistent with the data, usually leads to some interesting discussions in the processing of the activity. In practice, we usually end up agreeing on the assignment of grades, or our assignments are close enough that the difference does not significantly affect their grade on the activity.

I have found that this type of problem is very effective in introducing the students to the dynamics of a nonroutine problem. The essence of this problem is that the students are given some data for which they have no predetermined sense of the mathematical patterns involved, then by organizing and studying the data they "see" some fairly simple patterns and based on these patterns they can solve a problem. Notice that the students do not need to know or remember any predetermined formulas and that the students are constructing the mathematics to answer a relevant question. To reinforce the power of this process, I sometimes introduce the students to the formula for calculating the standard deviation and the clusters determined by the formula, and then comparing the results with their calculations without the formula, emphasizing the accuracy of their calculations without the formula and the difficulty of using the formula.

Many similar nonroutine problems can be generated by being sensitive to problems relevant to the students that can be answered fairly easily by generating data and looking for patterns. One type of problem is if there is a need to discriminate between the value of some items to a group. For example, suppose the class is planning a party and you need to know what food would be best received. It is fairly straightforward to have the class brainstorm eight to twenty items and then have each student vote for their favorite three foods. Usually the data, with some discussion or follow-up vote, will usually fairly clearly indicate the best choices. Typically, a few items will be clear favorites, some will be moderately liked and some will be clearly not preferred, final decisions will usually revolve around balance and cost.

Problem 9. "Five calculations": For this problem, each group of three students prepares for the following task: for each of 20 numbers, between 100 and 900, provided by the teacher (i.e., the specific numbers are unknown to the students before the task), reduce each number to 0 with five or fewer calculations involving addition, subtraction, multiplication and/or division by the numbers 1 to 9 only. For example, given the number 417, one solution would be: (a) subtract 1: $417 - 1 = 416$, (b) divide by 8: $416/8 = 52$, (c) add 4: $52 \times 4 = 56$, (d) divide by 7: $56/7 = 8$, and (e) subtract 8: $8 - 8 = 0$. For the task, each group is given two calculators and 20 minutes to complete the task. In addition to blank paper and pencil, each group may bring one 3 by 5 card with writing on one side. Groups are given one point if they reduce a given number to 0 in five calculations; two points if they reduce a given number to 0 in four calculations; and three points if they reduce a given number to 0 in three calculations. The groups have one to two weeks to prepare for the task. Depending on the class, I determine what I believe is the right balance between in-class and out of class planning time.

A discussion of one solution will clarify the nature of the problem. It turns out that there is a fairly straightforward method to generate a solution in five or fewer steps for the numbers 100–819: (a) if not divisible by 9, subtract a number less than 9 to result in a multiple of 9, (b) divide by 9, resulting in a number less than 91, (c) if not divisible by 9, subtract a number less than 9 to result in a multiple of 9, (d) divide by 9, resulting in a number less than 10, and (e) subtract to equal 0. For example, if given 784, then $784 - 1 = 783$; $783/9 = 87$; $87 - 6 = 81$; $81/9 = 9$; and $9 - 9 = 0$. The numbers from 819–900 are not as easy and require you to find numbers in the 819–900 range that allow two consecutive divisions by fairly large factors (less than or equal to 9 of course) or represent other similar special circumstances that allow reduction in four steps. Then you can generate an interval of solutions in five steps by first adding or subtracting to get that number. For example, the numbers from 855 to 873 can be reduced to 0 by first getting to 864 by adding or subtracting, then $864/9 = 96$, $96/8 = 12$, $12/6 = 2$, and $2 - 2 = 0$. It turns out that by this process you can find numbers that can be reduced to 0 in four steps that "allow" all numbers from 811 to 900 except 850 to 854 to be reduced in five steps. Finally, 850, 852, and 854 can be reduced by individual solutions (e.g., $854/7 = 122$, $122 - 2 = 120$, $120/6 = 20$, $20/5 = 4$, and $4 - 4 = 0$). There is no solution in five steps for 851 and 853. Investigations of situations that can be solved in less than five steps and consideration of how to work effectively as a group with limited resources and time will further improve a team's score and allow an opportunity for persistence. If a group discovers the solution in five steps for the numbers 100–810, there is plenty of room on the 3 by 5 card to write the solutions (using intervals) for 811–900, as well as information that facilitates solutions in three or four steps. I have had a number of groups that have scored the maximum number of points or within one point with the information they wrote on a 3 by 5 card.

From the point of view of mastering nonroutine problems, this problem is excellent for early in the curriculum. The problem requires few content prerequisite skills and no need to use written or people resources outside the group. In the directions I provide the students with the rules for divisibility (e.g., if the sum of the digits of the number is divisible by 9, the number is divisible by 9) and indicate that there is no more "mathematics content" from written sources that will be useful; that is, they need to develop their own methods. In addition, the problem lends itself naturally to a higher level of support by giving the groups one or two field-tests before the final test, providing a transition from trying something to persisting. For example, in practice, most groups in an average first or second year high school mathematics class believe they have developed a good strategy before the first field-test and wonder why a field-test is even necessary and are surprised to find that their strategies are not nearly as effective as expected. The structure of requiring a field-test provides the necessary support to help the students focus on the need for persistence in this problem. My hypothesis is that later in the curriculum the students would have naturally field-tested their strategy without the need of the teacher requiring or suggesting it. I will mention that I do not inform the students that there are two numbers that cannot be reduced in five steps and I do not include one of the two numbers that cannot be reduced on the field-test but include one on the final test. This practice provides an interesting twist to the problem, and a good basis for discussing persistence, since if one persisted with this problem in theory you would know that those two numbers were not possible.

Notice that this problem deals with content not considered particularly important in the academic high school curriculum. So, from the point of view of mastering content this task would not be considered very important; however, from the point of view of mastering nonroutine problems I have found this task to be excellent.

Reference

Caine, G., Caine, R. and Crowell, S. (1994). *Mindshifts: A brain-based process for restructuring schools and renewing education.* Tucson, AZ: Zephyr Press.

4

Geometry

Chapters 4 to 7 each contain a variety of nonroutine problems associated with specific mathematical content. The problems in each of these chapters are divided into three levels, "introductory," which are particularly appropriate for students with little or no experience solving nonroutine problems and usually have no significant content prerequisite skills needed for a solution; "intermediate," which are most appropriate for students that have some experience solving nonroutine problems and "advanced," which typically require a good amount of experience solving nonroutine problems, as well as some understanding of traditional mathematics content and/or success in academic mathematics courses. These three levels are somewhat dependent on your actual context and students and the levels are meant to give you some guidance concerning appropriateness as well as the amount of scaffolding you provide versus representing a definite level of difficulty.

Given space limitations, for most of the problems in these chapters, as well as Chapters 8 to 11, individual lesson plans will provide the information needed to implement the lesson in your context; however, the lesson plans will omit much of the detail discussed in Chapters 2 and 3. Most problems will include a sample student statement, as well as teaching suggestions which outline the needed directions to introduce the problem as well as suggestions for the processing, sample typical solutions, and, if appropriate, suggestions for enrichment for the problem. In some cases I use a slightly different format that I believe is more appropriate for a specific problem. In most cases, these suggestions will need to be modified to fit the specific context of your

DOI: 10.4324/9781003393283-4

classroom and school! Hopefully, the suggestions will give you a basis to determine the most appropriate implementation for your classroom.

Finally, the chapters from 4 on assume a reading of Chapters 2 and 3 which discuss in detail the theory underlying this approach, the pedagogy recommended and an in-depth discussion of ten introductory nonroutine problems. In summary, the directions for the many problems discussed in these chapters are less detailed than the problems in Chapter 3 but hopefully provide enough detail that the reader can adjust the directions to fit her/his context, based on the content of Chapters 2 and 3, as well as your own experience teaching.

The nonroutine problems discussed in this chapter focus on the content of an academic geometry course. The main purpose of these problems is to improve the student's ability to solve nonroutine problems, not to cover the traditional geometry course content. However, a few of the problems also explore or introduce content usually covered in a traditional geometry course, while others explore what might be considered enrichment topics in geometry not typically integrated into a geometry course.

Introductory nonroutine problems

There are five introductory nonroutine problems discussed in detail in Chapter 3 that concern geometry:

(1) **Calculating the area of an irregular closed curve.** Perhaps the easiest of the nonroutine problems that almost every student can engage with, independent of their mathematics ability.
(2) **Maximum area.** This problem integrates well with the topic of area, particularly of a rectangle, as well as concretely demonstrating that a given perimeter does not determine the area of a rectangle. The processing provides a good focus on trying something, as well as persistence.
(3) **Maximum volume.** This nonroutine problem allows a deep focus on persistence based on the class's solutions, including deciding when you have persisted reasonably well. Also, it is a very hands-on, open-ended problem that the students enjoy.
(4) **Expensive tape.** This is my favorite introductory nonroutine problem for introducing the three steps, the processing typically demonstrates well the three steps and the connection to student solutions. In addition, indirectly the students are introduced to the concept

that the lengths of the three dimensions of a rectangular solid do not determine the volume of rectangular solid; specifically, that the sum of the girth and the length of a rectangular solid do not determine the volume of the rectangular solid.

(5) **Most pleasing rectangle.** This nonroutine problem dramatically demonstrates the power of the three steps in solving what seems like an insoluble problem through trying something and persistence. It is also a good introduction to the Golden Mean. This problem also is relevant to the content focus of prediction and estimation.

Intermediate level

There are five problems at this level of difficulty:

(1) **"One inch of rainfall."** This nonroutine problem asks students to calculate the volume of rain that would fall in their community if there was 1" of rain.

A sample student problem statement. The task for this problem is to calculate the volume of rain that would fall on your community if there was 1" of rain. The purpose of this task is to give you practice solving a nonroutine problem involving volume and using resources. Specifically:

1   You will calculate the volume of rain that would fall on this community if there was 1" of rain. You will give your answer in terms of gallons of water and at least one equivalency which will give us a sense of just how much water your answer represents (e.g., how many showers could be taken with the water?).

2   You will turn in a written report including how you arrived at your final solution.

3   Only a few students will be asked to orally explain their project but each student should be prepared to give an oral presentation.

Teaching suggestions. In terms of prerequisite skills, the students need to have some concept of volume and need to be able to use resources to make conversions and determine the size of their community. To start, go over the directions and answer any student questions. You might want to emphasize that they are to find an equivalency for their answer that will clarify just how much water is involved and is interesting. Since the example of how many showers would be involved is cited in the problem statement, it should be clear that it would be inappropriate to use that as an equivalency. You might suggest that they imagine they are preparing a news report and want to use an equivalency that will interest viewers. When the projects are complete, collect the written reports. If, in reviewing the reports, any oral presentations seem appropriate, it is probably best to schedule them before the general processing of the project.

For processing, besides whatever comments naturally come out of reviewing all the reports, it seems important to discuss their experience in locating resources to answer the problem. What difficulties did they run into? What worked well?

This is an excellent individual project, but it certainly can be an effective cooperative group task. The oral reports, if any, and the processing will require a portion of a class period. The project can be completed in about five school days. A good extension is to calculate how much rain would fall on your community if it rained 40 days and 40 nights. Notice that students would have to make some assumptions about the amount of rain per hour.

(2) **"Congruency in triangles":** This problem requires students to evaluate how many different triangles can be constructed when given between two, three or four bits of information about the triangle. The students are given 17 situations to evaluate. The problem introduces the students to the concept of congruency in triangles. This can be an individual or group project. An individual project would probably be appropriate only if your students have already completed some nonroutine problems. This is one of a few nonroutine problems that also serves as an excellent problem to introduce traditional geometry content, specifically congruency in triangles. In my experience this problem facilitated an excellent understanding of congruency in triangles, followed by more formal instruction in the acceptable basis for proving congruency (e.g., ASA; SSS), as well as unacceptable methods.

Below is a sample student problem statement:

The task for this problem is to identify how many possible different triangles can be constructed given certain information. Specifically:

1   You will investigate 17 situations (provided in TRIANGLES: STUDENT WORKSHEET) for which you are given two, three or four bits of information about a triangle and you are asked to determine whether one, two or an infinite number of different triangles can be constructed consistent with the given information.

2   You will turn in a written report including your answers, a picture for each situation supporting your answer, and a description of how you arrived at your solutions.

3   Only a few students will be asked to orally explain their project, but each student should be prepared to give an oral presentation.

Teaching suggestions. This is an individual project; however, if your class has little experience solving nonroutine problems, it may be appropriate to assign this as a cooperative group problem. Note: For this problem AB = 5″ means that the length of line segment AB is 5″ (versus the line through the points A and B).

The task can be introduced by you doing four demonstration constructions (see the student worksheet and Figure 4.1): (a) AB = 4″, BC = 5″, AC = 7″, (b) m($\angle$A) = 50, AC = 5″, (c) m($\angle$A) = 50, AB = 5″, BC = 4″ and (d) m($\angle$A) = 50, m($\angle$B) = 60, m($\angle$C) = 70, AC = 4″. The first construction demonstrates conditions which allow only one triangle to be constructed, the second allows an infinite number, the third allows exactly two and the fourth allows one triangle (it should be clear from the fourth construction that one of the angles is not necessary to complete the construction).

Give the students the worksheet TRIANGLES: STUDENT WORKSHEET. Go over the directions and answer any student questions. The students require compasses, rulers and protractors to do the constructions. However, you should let students know that many, if not all, of the problems can be solved without compasses, rulers and protractors; that is, sketching based on reasonable estimation of measurements should be adequate. Also, it is not necessary that their pictures be exact in measurements as long as the picture demonstrates accurately the number of triangles that can be constructed and how those triangles can be constructed. You can suggest that students check with you if they are not sure if a picture is adequate.

You might want to emphasize that the four demonstration problems are representative of the types of solutions they will get for the 17 problems. Also, make sure students realize that for solutions for which an infinite number of triangles can be constructed, they need to show only three. The group activity requires at least a portion of three class periods with students working on the constructions outside of class also. The students require compasses, rulers and protractors to do the constructions. If time permits, each group can give an oral presentation on their results. The processing and follow-up of the activity is an appropriate time to introduce the methods for proving congruency of triangles.

This is a good problem to have students turn in a progress report half-way through the project. When the projects are complete, collect the written reports. If, in reviewing the reports, any oral presentations seem appropriate, it is probably best to schedule them before the general processing of the project.

Processing. Besides whatever comments naturally come out of reviewing all the reports, it seems important in the processing to discuss the role of persistence in this problem. You might even want to alert the students to include any examples of persisting in their written reports. This project naturally leads to a discussion of the rules of congruency in triangles.

Enrichment and Extension. This problem is very appropriate before introducing the formal rules for determining that two triangles are congruent. Also, a discussion of when SSA is a valid reason for congruency (e.g., HL in a right triangle) could be interesting. In addition, the nonroutine problem on congruency in quadrilaterals is a good extension of this project, especially for excellent mathematics students.

Triangles: student worksheet

Directions: For each example below you are to answer the following question: With the given conditions, can you construct exactly one, two or an infinite number of noncongruent triangles? In answering the above question construct, respectively, one, two or three triangles.

(1) AB = 3″, BC = 4″
(2) m($\angle$A) = 50, BC = 3″
(3) m($\angle$B) = 60, m($\angle$C) = 50
(4) AC = 6″, BC = 5″
(5) m($\angle$C) = 50, AB = 4″
(6) m($\angle$A) = 70, m($\angle$C) = 50
(7) AB = 4″, BC = 5″, AC =6″
(8) m($\angle$A) = 30, AB = 6″, BC = 3″
(9) AB = 5″, m($\angle$A) = 40, AC = 7″

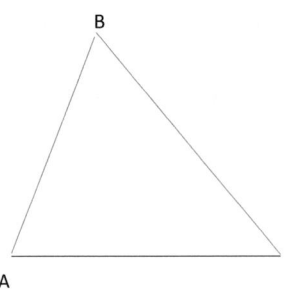

Figure 4.1 Reference triangle.

(10) m(∠A) = 50, m(∠B) = 70, m(∠C) = 60
(11) m(∠A) = 50, AC = 5″, m(∠C) = 60
(12) m(∠A) = 40, AB = 5″, BC = 4″
(13) m(∠A) = 40, m(∠C) = 60, BC = 5″
(14) m(∠A) = 90, AB = 3″, AC = 4″, BC = 5″
(15) m(∠A) = 60, m(∠B) = 40, m(∠C) = 80, AC = 5″
(16) m(∠A) = 60, AB = 4″, BC = 4″, AC = 4″
(17) m(∠A) = 50, m(∠C) = 80, AC = 5″, BC = 5″

Hint: Remember the four demonstration constructions: (1) AB = 4″, BC = 5″, AC = 7″, (2) m(∠A) = 50, AC = 5″, (3) m(∠A) = 50, AB = 5″, BC = 4″, and (4) m(∠A) = 50, m(∠B) = 60, m(∠C) = 70, AC = 4″.

Key to solutions: (1) examples 1 to 6 and 10 allow an infinite number of noncongruent constructions; (2) examples 7, 9, 11 and 13 to 17 allow exactly one construction and (3) examples 8 and 12 allow exactly two constructions.

(3) **"M & M's."** Students calculate how many M & M's it would take to fill their classroom. This problem also is relevant to the content focus of prediction and estimation.

Sample student problem statement: The task for this problem is to calculate how many M & M's it would take to fill this classroom. Specifically:

(1) Each group will determine at least three different ways to solve this problem and determine one answer as your best estimate.
(2) Each group will hand in a written report including your final estimation, an explanation of how you arrived at that estimation and why you believe the estimation is accurate. In addition, you need to document all the methods you considered (minimum of two additional methods) and any other factors that affected your solution.
(3) Each group will prepare an oral presentation of two minutes or less on your work.

Teaching suggestions. The only prerequisite skill is that students need to understand the concept of volume as differentiated from area. Give the students the problem statement for M & M'S. Go over the directions and try to answer any questions. It might help to emphasize that a major portion of the problem is to come up with a variety of methods; that is, a group that finds one method that is very accurate will not do as well as a group that finds several good to very good methods. You might want to suggest to groups that they brainstorm methods in their group first and then start to evaluate the quality of the methods. If possible and appropriate, allow times outside class that students can "inspect" the classroom. You need to be prepared for questions concerning what counts (e.g., the space desks occupy).

The students work in groups of three and are given a portion of two periods to work on their project. The oral reports and the processing will each require a portion of a class period. The project can be completed in five to ten school days. When the projects are complete, have the groups give their oral presentations. A good focus for the processing would be for the students to appreciate the wide variety of good methods to estimate the number of M & M's. Therefore, it would be appropriate to select one or two good methods from each group to discuss. Another focus might be what assumptions caused some of the methods to result in poor estimations.

Note: The requirement to use at least three methods to estimate is meant to encourage the step of persistence in arriving at their best estimate. One disadvantage of this approach is that students might identify three strategies to satisfy the requirement but miss the essence of using three or more strategies to facilitate an excellent estimate; e.g., they just average the three estimates, not usually a valid method. Therefore, one good focus of the processing could be whether groups effectively implemented three methods; e.g., did the group determine a good estimate based on a well-justified assessment of the methods?

Enrichment and Extension. A good similar problem is to estimate how far a trillion M & M's would stretch out if lined up in a straight line. Of course, any complicated estimation would be appropriate, including some of the problems in Chapter 6.

(4) **"Geometric constructions."** For this problem, the students prepare for a number of outdoor constructions by preparing "measuring devices" in the classroom, followed by actual constructions outside.

Student problem statement: The task for this problem is for your group to construct certain geometric figures using only measuring devices you develop in class. In preparing for the constructions, the only information you will have is that the constructions will be polygons and no side will be greater than 35 feet. The purpose of this task is to give you practice solving a nonroutine problem involving geometry and careful planning. Specifically:

(1) Your group will have two planning sessions to prepare for some geometric constructions I will give the class. The only information your group will have before I give you the actual problems is that the constructions will be polygons, no sides will be longer than 35', the

dimensions of the sides will be in feet (no inches) and some figures will have right angles.

(2) During one of your planning sessions, your group will construct measuring devices out of 20' of string I give your group. Rulers, scissors, and masking tape will be available for making the measuring devices. You cannot mark your measuring devices except for total length. For example, you cannot mark each foot on a ten-foot piece of string.

(3) During the actual constructions, your group will only have your measuring devices, scissors, nails for staking, and a supply of string (e.g., no rulers). For the actual constructions, we will go outside and each group will have a space to do their constructions. I will give one construction at a time and I will have a way to check the accuracy of each construction.

(4) Your group will have a portion of one class to prepare for the oral presentations. Each group will be asked to explain your strategy for at least one of the constructions. I will select the member of your group to present the oral report. There will be no written report for this project.

Teaching suggestions. Prior to the task students will need to understand what it is you are asking them to construct. For example, one construction is a right triangle in which they are given the hypotenuse and a leg. If you use that terminology obviously, the students need to know the definition of a right triangle, hypotenuse and leg. Of course, it is possible to give the same construction with a different description; e.g., construct a triangle with a 90 degree angle, the longest side 26', and one of the other sides 24'. You will notice that some problems, such as the right triangle construction, are easier if you remember certain properties (e.g., how to determine the third side of a right triangle). One option is to give the students a summary of the relevant properties of polygons and perhaps a few irrelevant properties to use during the constructions and planning. The problems are challenging with or without a summary.

Go over the directions and answer any student questions. Make sure that students have a sense of how the constructions will work, including the fact that they will be limited in the materials available to them for the constructions and that they will not know what constructions they will be required to do until the actual time that they are given the problems. Also, the groups need to understand that the measuring devices cannot be marked except for total length (e.g., a good solution might be to construct a 10', 5', 2' and 1' length).

The problems are meant to be constructed outside using string and nails to construct the sides and vertices. However, the constructions can just as easily be done inside in a large room such as a gymnasium with a few adjustments (you would not want to use nails!). Depending on the amount of time available, you can use all or some of the following constructions (a scoring technique based on starting with 100 or 200 points is included in parentheses) (a) A triangle with sides 15', 17' and 22' (score: one point deducted for each inch off each side), (b) a rectangle 12' by 16' (score: one point deducted for each inch off each side and two points off for each inch off one diagonal which should be 20'), (c) a right triangle with a hypotenuse of 26' and a leg of 24' (score: one point deducted for each inch off each of the given sides and two points off for each inch off the third side that should be 10'; one point off for each degree off from 90 degrees), (d) an isosceles trapezoid with bases 10' and 34', and legs of 20' (score: one point deducted for each inch off each of the given sides and two points off for each inch off the altitude which should be 16') and (e) a regular hexagon with 15' sides (score: one point deducted for each inch off each of the sides and two points off for each inch off one of the diagonals which should be 30'). Also, they can be asked to construct a triangle of sides 13', 15' and 30' with the instructions that they cannot ask any questions and when they have their answer to let you know (the triangle cannot be constructed because the sum of 13' and 15' is less than 30'). When the constructions are complete give the groups time to prepare their oral presentations. Have each group explain one or two of their solutions.

Processing. Besides whatever comments naturally result from reviewing all the reports, discuss how the groups prepared for the constructions. One question might be what groups would have done differently if given the same type of task again. This task emphasizes group skills, especially during the actual constructions; therefore, a discussion of how groups functioned during the constructions would be appropriate.

Suggested format. The students work in groups of three and are given a portion of two periods to develop their measuring devices and plan, a class for the actual constructions and a portion of a period to prepare their oral report. The oral reports and the processing will each require a portion of a class period. The project can be completed in six to ten school days. Depending on your classes' experience, you may want to include a practice round in which they complete one construction. The processing of the practice round can be followed by an additional group planning session in preparation for a second more difficult round.

Enrichment and Extension. This is a good task to give a second round of new constructions (e.g., rhombus with given diagonals, quadrilateral given the four sides or a parallelogram given the sides and a diagonal). A comparison of the first round and second round could be very valuable.

(5) **"Quadrilaterals."** Students discover as many theorems as they can for a variety of quadrilaterals. This nonroutine problem requires the students to experiment with software that allows them to easily construct a variety of quadrilaterals and measure the angles and the lengths of various parts of the quadrilateral, thereby looking for patterns. The evaluation of their work includes an assessment of the quantity of theorems, accuracy of the theorems (Is the theorem true? Is it stated for the least restricted type of quadrilateral?) and quality of the theorems (includes the uniqueness of the theorem and power of the theorem). The students use computer software that allows the student to easily construct quadrilaterals and generate data.

Student problem statement. Your group will develop a list of theorems that are true for quadrilaterals or special types of quadrilaterals. An acceptable theorem for this project is any statement that is true for all quadrilaterals or for all quadrilaterals of a certain type (e.g., rhombus, square). Theorems can include, but are not limited to, statements about (a) angles and relationships among angles (e.g., equality of angles, sums of measures of angles or size of angles), (b) sides and relationships among sides, (c) area or perimeter, (d) diagonals and (e) constructed line segments with special properties such as medians, angle bisectors and perpendiculars. Theorems will be recorded on separate sheets labelled quadrilaterals, parallelograms, rhombuses, rectangles, squares and trapezoids. Theorems should be recorded on the least restrictive list. For example, a theorem true for all quadrilaterals and for parallelograms should be recorded on the quadrilateral list. A theorem true for rectangles and parallelograms, but not true for all quadrilaterals should be recorded on the parallelogram list.

Teaching suggestions. This is an example of a nonroutine problem for which the level of support you provide the students is critical to whether they improve their ability to solve nonroutine problems. Specifically, the students need enough help to have a sense of how to generate data using the computer and how to work with the data. One way this can be accomplished is by giving students a variety of situations to construct and look for patterns. For example, I have found that the investigation of the following four examples is usually adequate to prepare average students for the more open-ended version (a) Is there a pattern involving two angles in a trapezoid on the same transversal of parallel lines (a picture is provided for the students)? (b) Is there a pattern involving the diagonals of a rectangle? (c) Is there a pattern involving any of the four angles of a rhombus? and (d) What conclusions can you draw concerning the diagonals of a rhombus? Of course, the teacher needs to be sensitive to the optimal point where the students have enough experience investigating quadrilaterals so that the problem will not be frustrating, yet not enough experience to make the problem straightforward and thereby not useful for the objective of mastering nonroutine problems. The key point is that the objective of mastering nonroutine problems is what determines the structuring of the support.

A similar nonroutine problem focusing on triangles is appropriate either before or after this problem on quadrilaterals.

Advanced

There are two problems at this level of difficulty:

(1) **"Approximating π"**: This problem is also relevant to the content focus of number theory. The task for this problem is for each group to approximate π using ten different methods. Five of these methods must be labeled: a method that does not involve polygons, a method that uses trig but does not use trig tables or a calculator, a method without trig, a method without a calculator and your best method. In addition to evaluating the quality of the five required methods, the ten methods, as a group, are evaluated on three criteria: originality and creativity, variety of methods and the quality and accuracy of the calculations. The variety of criteria help insure that the students will use a variety of techniques.

Student problem statement: The task for this problem is for your group to approximate π using ten different methods. The purpose of this task is to give you practice solving a nonroutine problem involving evaluating alternatives and careful calculations. Specifically:

1   Your group is to approximate π using ten different methods. For each method, you must include your data and describe the instruments you used. Five of these methods must be labeled: a method which does not involve polygons, a method which uses trig but does not use trig tables or a calculator, a method without trig, a method without a calculator and your best method.

2   In addition to evaluating the quality of the five required methods, your ten methods as a group will be evaluated on three criteria: originality and creativity, variety of methods and the quality and accuracy of the calculations.

3   Each group will prepare an oral presentation of two minutes or less on no more than three of your methods. I will select the member of your group to present the oral report.

4   Your group will turn in a written report. For this project, the written report will be restricted to identifying your ten methods and documenting how you calculated each method. Your report should start by documenting the five required categories, clearly labeling each category.

Teaching suggestions. A major strength of this problem is that it forces the student to evaluate potential solutions. The student must decide on the ten methods which together will result in the highest possible grade on the project. Certainly, there are a number of different sets of directions that could be given to encourage this emphasis on evaluative skills. Here I will outline one set of such directions. Of course, the reader is encouraged to adjust the directions to fit your context. To introduce the assignment you might give the students an example of approximating π such as the following: If you have a square inscribed in a circle with the radius equal to one unit, then the side of the square is equal to the square root of two, the circumference of the circle is equal to 2π, and the perimeter of the square is equal to four times the square root of two. Then if we say that the circumference of the circle is approximately equal to the perimeter of the square, then

$$2\pi \cong 4\sqrt{2}$$

If we substitute 1.414 for the square root of two and solve for π we get that π is approximately 2.828.

Let students know that they may use formulas involving circumference, area, volume or physical forces (such as centrifugal force), but not infinite series such as $\pi/4 = 1 - 1/3 + 1/5 - 1/7 + \ldots$ or trig formulas such as $\pi/4 = 4 \arctan(1/5) - \arctan(1/239)$. You might encourage students to ask you individually about the appropriateness of a questionable method. The rationale for the above directions is to increase the likelihood that the student will use a variety of methods. Consequently, in discussing the directions, it could be useful to point out that the method used above to approximate π (obtaining $\pi = 2.828$) is not very accurate and that you could improve the method by increasing the number of sides. At the same time, make it clear that an assignment in which most of the methods involved polygons (even good approximations) would not receive a good grade, certainly losing points for originality and variety. Additionally, students should realize that the different categories for grading force them to evaluate the ten methods they include for balance, a very creative method may not be very accurate, etc.

Comment: Since students can easily look up as many digits of π as they need, it is legitimate to be concerned about students fudging data. The way I have dealt with this concern is by making it clear to students that part of the assignment is to include their data and how they obtained it. In addition, I make it clear that well carried out, creative methods will contribute to an excellent grade even if the method is not as accurate as other methods. These measures seem to effectively minimize fudging. This is a good project to require a progress report halfway through the project.

Suggested format. The students work in groups of three and are given a portion of three periods to work on the project and prepare their oral report. If appropriate for your context, this could be an excellent individual problem. The oral reports and the processing will each require a portion of a class period. The project can be completed in six to ten school days. When the projects are complete, have the groups give their oral presentations.

The work of two students will be contrasted to give a sense of the range of solutions for this problem. The first student received a fair grade and the second student received an excellent grade. For the best method the first student used the law of cosines with a 360-sided polygon inscribed in a circle to obtain a value of 3.1415513; the second student wrote a program for calculating the area of a quarter of the unit circle by rectangular approximation, using 100,000 rectangles the student got a value of 3.141592682 (the first eight digits are accurate). For the four other labeled methods, the first student (a) compared an inscribed circle with a square by weighing for a value of 3.08 (without trig), (b) measured circumference by rolling a circle on a ruler and solved for π for a value of 3.14 (without a polygon), (c) used the law of cosines with a 45 degree angle for a value of 3.06 (trig without charts) and (d) used an inscribed hexagon for a value of 3 (without calculator). The second student (a) used a cylinder, density formula, calipers and three trails for a value of 3.11 (without trig), (b) used a sphere and determined volume by displacement for a value of 3.1746 (without a polygon), (c) used an inscribed hexagon for a value of 3 (trig without charts) and (d) wrapped string around a circle ten times and used a caliper to measure the diameter for a value of 3.136 (without calculator).

For the five remaining methods the first student (a) used a pendulum for a value of 3.1474185, (b) used the law of cosines with an angle of five degrees for a value of 3.1406, (c) used a cylinder and displacement for four runs for a value of 3.13, (d) circumscribed a square around a circle for a value of 4 and (e) averaged the area of a circumscribed and inscribed triangle for a value of 3.8971. The second student (a) used the formula for centripetal force for a value of 3.144418905, (b) used the fact that the acceleration of gravity equals the acceleration of a pendulum to calculate π as 3.14128629, (c) inscribed a polygon with 2^{12} sides in a circle for a value of 3.141581267, (d) circumscribed a similar polygon for a value of 3.14159327 and (e) used a program to calculate the length of an arc equal to a quarter of a circle using 100,000 line segments for a value of 3.141638452.

Processing. Besides whatever comments naturally come out of reviewing all the reports, it seems important in the processing to select two methods from each group, one which includes a positive point in reference to the problem solving skills and one which includes an aspect that could be improved. One suggestion is that at least some of the methods you pick for improvement might focus on manipulative skills; the question for discussion could be how to improve the accuracy of the calculations. Could the quantity of measurements be increased, e.g., attempt more trials? What instruments were used and could they have been improved? Did the student provide for checking the accuracy of the calculations? For example, if we compare the solutions of the two students discussed above, the second student clearly demonstrated better manipulative skills. For example, for the method involving centripetal force the second student used a precision balance and ruler for measuring mass, weight and the radius and completed 20 revolutions to measure time with a stopwatch. Both students determined π using the diameter and circumference of a circle, the first student measured one revolution versus ten for the second student.

The third step of problem solving, persistence, might be a good focus for the discussion. You can have the students compare methods in which they persisted and felt satisfied versus methods they gave up on or "settled" for a less than adequate solution. Or you can ask each student to identify one method they did not include in the ten methods and use these as a basis to discuss persistence – how could these discarded methods have been used effectively?

This nonroutine problem is a good one to focus on the adequacy of documentation. You might pick out the best and the worst documented method for each student either for individual discussion (e.g., comments on the assignment) or for class discussion.

Enrichment and Extension. Perhaps the most famous approximation of π was by Archimedes in the book The Measurement of a Circle in which he makes the proposition that "The circumference of any circle exceeds three times its diameter by a part which is less than 1/7 but more than 10/71 of the diameter," i.e., $3\,1/7 > \pi > 3$ 10/71. This approximation is within .2% of the actual value. The technique that Archimedes used is of interest. He inscribed and circumscribed regular polygons with sides of 6, 12, 24, 48 and 96 sides. A hexagon is simple to inscribe by marking off six radii to locate the vertices. Then by constructing the tangents to the vertices you can construct the circumscribed hexagon. Then by bisecting subtended arcs the other polygons can be constructed. Then, by comparing the circumscribed polygon with the inscribed polygon he obtained his approximation. The practical limits of the mathematics of Archimedes' time made calculations beyond a polygon of 96 sides prohibitive.

(2) **Area under a curve.** This is an area problem that requires the students to determine the area bounded by $y = x^2 + 1$, $y = 0$, $x = 0$ and $x = 3$. In terms of content focus, this problem introduces the concept of integration in calculus and is particularly appropriate for a precalculus or calculus course. As mentioned in Chapters 1 and 2, I consistently assigned this problem in the beginning of my calculus classes, and, in fact, the solutions of my calculus students convinced me of the need for a curriculum of nonroutine problems, even for the best mathematics students. Therefore, I will discuss this problem in more detail in Chapter 5. However, given the pedagogy described for a curriculum of nonroutine problems, this problem can certainly be appropriate for any mathematics class in which students have the necessary content prerequisite skills (e.g., given a function $f(x)$, determine the value $y = f(x)$).

5

Algebra and precalculus

This chapter concerns nonroutine problems that involve mathematical content concerning algebra, precalculus and calculus. Only four of the problems in this chapter are at the introductory or intermediate level, while nine problems are at the advanced level, requiring not only experience solving some nonroutine problems, but also some understanding of mathematical content. Three problems focus on abstract thinking by requiring students to explore three functions based on the abstract definition of the function. For example, a logarithmic function is explored given the abstract property that $G(xy) = G(x) + G(y)$ and three values of the function.

Introductory nonroutine problems

There are two problems at this level of difficulty:

(1) **"Dividing a line segment."** The task for this problem is to develop a quick and accurate procedure to locate a point C on a line segment AB such that AB/AC = AC/BC. If the line segment AB is one unit, then the larger segment is the Golden Mean. Students are instructed that their procedure should work for any line segment and, as much as is possible, have the following characteristics: it is accurate, it is quick and it requires no or a few simple tools. The procedure is to be

DOI: 10.4324/9781003393283-5

written so that another person can follow the procedure. Accuracy and speed are tested by giving each procedure to a few students who use the procedure to divide a line segment AB (provided by the teacher) and record their starting and ending times. I have identified this nonroutine problem as introductory; however, the actual level of the problem depends on how the problem is introduced and the level of support provided. For example, at the level of algebra 1, this problem can be an excellent introduction to nonroutine problems by focusing on the step of "trying something" by substituting values for x until a specific value for x becomes clear. As an introductory problem, it is usually appropriate to demonstrate with the class what happens when you substitute a few values. Of course, the need to develop a quick and accurate procedure to locate the point C adds an additional dimension to the problem. Typically, as an introductory problem, it is most appropriate to be given as a cooperative group activity.

Without a good amount of teacher assistance this problem is appropriate only after the students have had experience with a variety of nonroutine problems or are excellent mathematics students. For example, students with experience with nonroutine problems have enough confidence in the step of trying something to do some experiments and quickly begin to notice some patterns (e.g., point C is always past the midpoint) which eventually lead them to an acceptable solution. In the typical solution, the student (or group of students) determines an approximation of a constant that allows the student to determine the length of one of the unknown segments, given the length of line segment AB (e.g., determine the length of line segment AC by dividing line segment AB by 1.6). The procedure typically consists of using a calculator and ruler to determine the location of point C. Depending upon the student's mathematical background and/or experience with solving nonroutine problems, the value of the determined constant ranges from rough approximations to accurate values determined by solving the appropriate quadratic equation or sophisticated trial and error.

Perhaps the most interesting solution of a student is the following:

… remove excess paper at the end of the segment, either with scissors, or by folding the material behind itself, and then fold the segment in half, then in half again, then in half one last time producing eighths. Point C should be made on the fifth crease in.

The student had determined an accurate value for the constant but argued that

Other methods involving the decimal would be more accurate [5/8 =.625 versus.618…] but would involve a calculator or a great deal of time. I feel that the method I presented was as simple and easy as one could expect, requiring no tools, no math, and little time. The discrepancy [from correct answer] is acceptable as considering the increased time required by other methods."

Many times the discussion and processing of this problem help students refine their abilities to develop clear and effective algorithms. For example, students developing the typical solution involving determining a constant see how other students with similar solutions determined their constant and wrote their procedures. In addition, some students get very specific feedback. For example, one student developed a nice procedure of using a protractor and compass to determine the point C, using the fact that a 52-degree angle has a cosine approximately equal to .618. However, the second step of the student's three-step procedure was unclear, "Measure a 52-degree angle, with the protractor, leaning into the segment, and from the end of the segment where the arc is." Through feedback from another student, the student was able to write a much clearer step. In his evaluation of what he learned working on this problem, he indicated the value of the feedback in helping him improve his ability to write clear procedures.

Solutions to the problem are somewhat effected by the students' mastery of mathematics content. In the above example, the student had already studied trig functions and used the cosine of 52 degrees for the solution, a solution not readily available to a student who has not studied trig. You might also notice that if you label the length of line segment AC as x and if the length of AB is 1, then x is the solution of the equation, $1/x = x/(1-x)$ which can be solved as a quadratic equation. However, few students both realize that it can be simplified as a quadratic equation and also remember how to solve a quadratic equation by the formula. For example, even in my calculus classes perhaps one or two students eventually realize they can solve the equation that way.

As is typical when students invest a good amount of time in solving a nonroutine problem, students appreciate a discussion of other solutions by mathematicians and the mathematics involved much more than if the topic is introduced without the experience of solving the nonroutine problem. Note also that this problem is good before or after the nonroutine problem, "Most pleasing rectangle" (described in Chapter 3) since both involve the Golden Mean.

(2) **"Graphing 1 and 2":** The introductory nonroutine problem "Graphing 1" is discussed in detail in Chapter 3. "Graphing 2"includes some more difficult graphs such as graphs with asymptotes and undefined points. It is meant to be given after students have completed "Graphing 1," perhaps later in the year, if you believe the students are ready for a more challenging nonroutine problem. It is also a good nonroutine problem for more advanced mathematics courses such as Algebra 2 and Precalculus. Below are the directions for "Graphing 2."

Sample student problem statement: The task for this problem is to determine the graph of the solution set of an equation given the equation. The graphs require you to try some values, look for patterns, and persist until the entire graph is clear. Specifically:

1 Your group will graph five equations. To graph each equation, you will need to determine some easy points, look for patterns, select appropriate sections to determine additional points and repeat this process until the entire graph is clear.

2 Each member of your group will be prepared to present an oral presentation of two minutes or less on each of your graphs. I will select the graph and the member of your group to explain the graph in the oral report.

3 In addition, your group will turn in a written report including a graph for each equation, a record of the sequence of points you determined for each graph and an explanation of why you chose the sequence of points.

Teaching suggestions. Prior to working on the problem students need to understand the connection between an equation and the graph of the points that satisfy the equation and how to plot points on an x-y coordinate system. In addition, students will need to understand specifically how to determine points that satisfy certain types of equations (e.g., equations solved for x such as $x = y^2$). Most classes also need a good introduction to the process of using the three steps to graph before starting the problem. For example, the following process can be suggested to students: try something by substituting some easy values (e.g., $x = 0, 1, 2, 3$ and 4); look for patterns indicating sections of the graph that seem clear and sections that do not seem clear; try additional values to check the clear sections and to investigate the unclear sections and repeat (or persist) until the entire graph is clear. The following examples can be used as demonstrations:

a) $y = x^4 + 1$: Start by substituting $x = 0, 1, 2, 3$ and 4 and solving for y (the scale for the y axis needs to be adjusted). Ask the students what sections of the graph seem clear $(x > 0)$ and which sections need more investigation $(x < 0)$. You can ask the students to predict the part of the graph for $x < 0$ (usually you will get about three different predictions that visually make sense). Try additional values such as $x = 2.5$ and $x = 10$ (to verify the clear parts of the graph) and $x = -1, -2, -4$ and -10 until the entire graph is clear. In reviewing the process, point out that the number of points needed to be clear about the graph will vary with the student and the equation. For example, some students may not need to substitute any values of x less than 0 because they realize that the graph of $y = x^4 + 1$ has to be symmetric to the y-axis; others may require quite a few values before they feel confident that their graph is correct. What needs to be emphasized is that you substitute values until you are confident that the graph is correct. That is the essence of the third step of persistence for this problem.

b) $x = y^2 + 4y + 2$: Demonstrate similarly, emphasizing the strategy of substituting for y instead of x in this type of equation.

c) $y = 12/x$: The demonstration for this equation should include a discussion of the two scenarios that can occur at an undefined point when the denominator equals 0; that is, a vertical asymptote or a continuous graph except at the undefined point (graphed by putting an open circle at the undefined point). Students should realize the need to try nonintegral values close to the undefined point. Let the students know that if an undefined point where the denominator equals 0 occurs in one of their group's problems that they should realize that one of the two above options will occur. You might add the example of graphing $y = (x^2-9)/(x+3)$, with an undefined point at $x = -3$.

d) $x^2 + y^2 = 25$: the demonstration for this equation needs to emphasize how to determine points for an equation for which substitution for either x or y does not always lead to a simple calculation of the second variable. You might need to include a discussion of the fact that the graph of $x^2 + y^2 = 25$ is not the same as the graph of $x + y = 5$ (the "square root" of both sides). The demonstration should give students a sense of how to "spot" easy values to substitute (e.g., x = 0 or y = 0) and how to determine values that are not rational. In addition, students need to understand that equations such as $y^2 = 25$ have two solutions.

The purpose of the above demonstrations is to give the students the assistance they need to make the problem at the appropriate level of difficulty. Without the demonstration most students would find the problem frustrating. Students generally need both a sense of how to apply the steps in this problem and an understanding of some of the specific methods for trying something for certain types of equations (e.g., graphs with asymptotes, functions of y). However, in order to keep the problem at the appropriate level of difficulty, you should avoid discussing general features of types of graphs such as the relationship between the graph of f(x) and |f(x)|, the shape of even and odd functions and the relationship between the graph of f(x) and its inverse.

"Graphing 2" is a powerful demonstration for students of using the three steps to solve a difficult problem. They are given equations to graph which probably seem very difficult or impossible to them but, by applying the three steps, the graphs "miraculously" gradually become clear. However, the success of "Graphing 2" depends upon the students' understanding of how to apply the steps. Therefore, for most classes it is appropriate to have the students take notes on the four examples and to give a quiz on the examples (see the Graphing quiz below). Of course, experience completing "Graphing 1" provides a good basis for completing "Graphing 2."

Go over the directions and answer any student questions. You might want to emphasize that each group will be responsible for only five of the problems and that they will need to apply the three steps as demonstrated to solve these problems. Each group should be assigned five problems, one each from the following intervals: problems 1 to 4, problems 5 to 8, problems 9 to 12, problems 13 to 16 and problems 17 to 20. Problems 21 to 24 are extra credit.

This problem is meant to be completed without the use of graphing calculators. Problems later in the curriculum assume the use of a graphing calculator; however, the use of a graphing calculator for this problem would interfere with the major objective of the curriculum which is to improve the student's ability to solve nonroutine problems. Therefore, you need to take measures to insure that graphing calculators are not used (see directions for "Graphing 1").

In addition to the demonstration of the four examples, it might be appropriate either to require progress reports or, if the initial solutions of a group are not adequate, to require redoing of the graphs with additional guidance. In any case, you need to provide a structure that will result in the great majority, if not all, of the students successfully experiencing the three steps, meaning most groups would correctly graph at least three out of five graphs. At certain points, I do check students' points for accuracy, meaning that they are actual points on the graph; however, I do not tell them if their total graph is accurate. Later in the curriculum, I would expect them to discover inaccurate points without my assistance.

When the projects are complete, have the groups give their oral presentations. The amount of time you are willing to allow for the oral reports will determine how many of the examples are discussed in the oral reports. In any case, a good procedure is to have a different student discuss each of the graphs.

Processing. Besides whatever comments naturally come out of reviewing all the reports, it seems important in the processing to emphasize the step of persistence, especially the process of trying values until the graph is clear. For example, many students at some point in this problem will ask how many points are necessary to determine each graph, as if there would be one fixed answer. By the end of the processing, most students should understand that you need to persist until it is clear that you have an accurate graph and the number of points needed will vary with each different student. Therefore, in the discussion, it is valuable to focus on both examples and nonexamples of persisting. For graphs that were not completed correctly, it could be valuable to graph just the points determined for the graph and asking the class if there are additional sections that need points and then completing the graph together.

Suggested Format. The students work in groups of three and are given a portion of two periods to work on the project and a portion of one period to prepare their oral report. The oral reports and the processing will each require a portion of a class period. The project can be completed in six to ten school days. When the processing is completed, give the students the attached Graphing problems, quiz. The average student should be able to graph at least two of the graphs correctly.

Enrichment and Extension. Graphing 2 includes some more difficult types of graphs that can be discussed such as graphs 21 to 24.

Graphing problems

Graph the solution set of the following equations:

(1) $y = x^2 + 3$
(2) $y = x^3 + 2$
(3) $y = 3x^2$

    (4) $y = 2x^2 - 1$
    (5) $y = x^2 + 6x + 3$
    (6) $y = x^2 + 3x - 1$
    (7) $y = x^4 - 2x^2$
    (8) $y = x^3 - x$
    (9) $x = y^3$
  (10) $x = y^2 - 2$
  (11) $y = 2^x$
  (12) $x = 2^y$
  (13) $x^2 + 4y^2 = 16$
  (14) $4x^2 + 9y^2 = 36$
  (15) $4x^2 - 9y^2 = 36$
  (16) $4y^2 - 9x^2 = 36$
  (17) $y = 1/(x - 5)$
  (18) $y = x/(x - 3)$
  (19) $y = x/(x^2 - 9)$
  (20) $y = (x^2 - 9)/(x - 3)$
  (21) $y = 5 - x^2/3$
  (22) $y = x^2/(x^2 - 9)$
  (23) $x^3 + y^2 = 8$
  (24) $y = x^2 + 5x + 6\,|$

Graphing quiz

Graph the solution set of the following equations:

  (1) $y = x^4 + 1$
  (2) $y = 12/x$
  (3) $x = y^2 + 4y + 2$
  (4) $x^2 + y^2 = 25$

Intermediate level

There are two problems at this level of difficulty:

  (1) **"Axioms of algebra":** Students generate a list of algebraic statements that are always true. The project is evaluated on completeness and conciseness.

      Sample student problem statement: The task for this problem is for your group to generate a list of equations that are always true. For example, the equation: $x + 3 = 3 + x$ is always true; that is, for every number you substitute for x, the equation is true. In contrast, the equation: x + 3 = 10 is only true when x = 7. Specifically:

1   Your group will turn in a list of equations and documentation indicating how the list was developed, including all equations that were considered and how equations were combined.

2   You want your list to be both comprehensive (including as many equations as possible) and concise (as few equations as possible without losing comprehensiveness).

Teaching suggestions. Prior to starting, students need to have an understanding of what it means for an equation to be true and what it means for an equation to be always true. Be sure that the students have a reasonably clear picture of what is required. Go over the directions and examples with them, making sure that they understand what it means for the list to be comprehensive and concise. It might be helpful to encourage groups to check with you during the project if they need clarification concerning combining equations for conciseness or other questions concerning the appropriateness of their list. Be sure that students understand that a group of similar equations expressed as a pattern (e.g., x + 1 = 1 + x, x + 2 = 2 + x, x + 3 = 3 + x, ….) is not as concise compared to one equation such as x + y = y + x. It may be best to limit the use of variables to just three (e.g., x, y and z) to make the lists more manageable.

Processing. A primary reason for this problem is its relevance to a student's first in-depth exposure to algebra. Therefore, the processing might focus on developing a class list that will be referred to throughout the year. In compiling a list, try to include at least one equation from each group. Obviously, you will want your list to include all the important axioms of algebra; however, for instructional reasons it may not be appropriate to make your list as short as is theoretical possible. For example, an equation such as:

$$(x+a)(x+b) = x^2 + (a+b)x + ab$$

might be appropriate, even though it can be proven from other axioms. Of course, when it seems appropriate in the course, you might want to show students that it can be derived from the other equations on their lists. A second focus of the processing might be to help the students develop an appreciation of the process of axiomatizing. Sharing relevant examples from each group's documentation is one way of focusing attention on axiomatizing. For example, if one group combined the equations 1 + x = x + 1, 2 + x = x + 2, 3 + x = x + 3,…

into x + y = y + x, that would be an appropriate example for discussion.

Suggested format. The students work in groups of three and are given a portion of two periods to develop their list. The processing should require a portion of a class period. The project can be completed in four to eight school days.

Enrichment and Extension. Students can investigate what can be done to an equation that results in a new equivalent equation. Students can be encouraged to include methods that work in only special circumstances or that have exceptions, as long as the qualifications and exceptions are documented (e.g., dividing both sides of an equation by the same number, except by zero; squaring both sides of an equation, as long as the roots are checked). An interesting enrichment topic is a discussion of Euclid's parallel axiom. The axiom basically states that if one is given a straight line and a point not on the straight line, exactly one straight line can be drawn through the point parallel to the given straight line. This axiom is important for a number of reasons. First, you cannot prove that the sum of the measures of the three interior angles of a triangle equal 180 degrees without that axiom. Second, for hundreds of years one of the most significant problems in mathematics was whether the axiom was necessary; that is, could the axiom be proved from the other axioms of Euclidean geometry? Historically, a variety of proofs were presented as valid only until proved not valid by another mathematician. Finally, a mathematician trying to disprove such a proof realized that, in fact, the axiom was necessary for Euclidean geometry and could not be proved from the other axioms and two alternative axioms resulted in two completely different geometries. For example, the alternative axiom that given a straight line and a point not on the straight line, no straight line can be drawn through the point parallel to the given straight line results in a non-Euclidean geometry in which the sum of the measures of the three angles of a triangle is always more than 180 degrees. This is easily illustrated by a triangle on the surface of a globe (sphere) that has one vertex starting at the North Pole and two vertices on the equator. The two base angles are right angles making a sum of 180 degrees and the third angle makes the sum of the measures of the three angles greater than 180 degrees. Before sharing this information, I like to ask the students what the sum of the measures of the three angles of a triangle equals. When they answer 180 degrees, I say not for all triangles and I ask the students to find a triangle with more than 180 degrees.

I have a globe in the room and let them know that such a triangle can be found in the room.

(2) **"Movement and graphs":** This problem requires students to determine how to move in such a way that the graph of the relationship between time and distance traveled will correspond to certain given graphs. The purpose of this task is to give students practice solving a nonroutine problem involving formulating experiments to gather data and persisting until finding an appropriate experiment.

   Sample student problem statement: The task for this problem is to determine how to move in such a way that the graph of the relationship between time and distance traveled will correspond to certain given graphs. The purpose of this task is to give you practice solving a nonroutine problem involving formulating experiments to gather data and persisting until finding an appropriate experiment. Specifically:

   1   Your group will try to figure out how to move in such a way as to correspond to the following graphs plotting time (t) and distance (D) in feet from a probe: $D = 2$; $D = 2t$; $D = t^2$; $D = 2^t$; and $D = \log_2 t$.

   2   Your group will be able to experiment with the probe in class but you will also need to think about the problem outside of class so that you can use your limited class time well. When your group believes you have a good match for one of the graphs, you will select five of the data points to turn in for that graph and show your complete graph to the teacher.

   3   In addition, for each graph, one member of your group (your choice) may be required to model the movement and each member needs to be able to model at least one of the graphs other than $D = 2$.

   4   Your group will turn in a written report describing your work including documentation of each experiment you tried and why you tried the experiment.

   5   Each group will be required to model at least one of the graphs using the probe. I will select the graph(s) your group will try to model.

Teaching suggestions. Prior to starting the task students need to have experience with graphing, including at least some experience with nonlinear graphs such as the nonroutine problems "graphing 1" and "graphing 2." The students need to understand how to determine points for the graphs, particularly $D = 2^t$ and $D = \log_2 t$. The equation $D = \log_2 t$ can be written as $t = 2^D$

if students are not familiar with log notation. The equations they work with can be adjusted to be consistent with your class's experience. For example, $D = \log_2 t$ may be inappropriate for some classes. In addition, students need to be comfortable with whatever technology you are going to use to gather and analyze the data. Go over the directions and answer any student questions. With most classes, it will be appropriate to demonstrate the technology to insure the familiarity with the tools they will need to complete their experiments. For example, you may have them try to move in such a way as to create the graph of a straight line (any straight line), and then to move in such a way as to create a graph which is not a straight line. If the class has had little experience with probes, you may want to do the graph $D = 2t$ together, alternating between class discussion or presentation and working in small groups with the probes. What is essential is that the students are comfortable enough with the technology so that the probes can be used effectively as a tool to gather data.

You might want to emphasize the need for the students to think and experiment outside of class time given the limited time they will have to experiment in class. If possible, allow for experimenting with the probes outside of class time. In any case, the in-class time limitations provide a good focus for the group skill for this task. Specifically, I would suggest that the evaluation of the group skill for this nonroutine problem should focus on the following question, "How well did your group use your limited class time? Was your group prepared to effectively use its resources and time during class time?" The criteria for evaluation could be your observations during class and the group's written summary of how they worked on the group task. As a class, you could brainstorm how a group demonstrating (and not demonstrating) the group skill might look to an observer. In addition, you can brainstorm what a group might do outside of class to insure a good use of their class time.

You need to make sure that the problem can work well with the technology you have available. For example, students will not appreciate the difference between the graphs of $D = t^2$ and $D = 2^t$ unless they can gather data for distances greater than 30 units (less than that distance the graphs look very similar). Therefore, you might need to adjust the units you use for measurement. In any case, it is important that you attempt the experiment yourself before you give it to the students.

When the projects are complete, have at least one person from each group demonstrate one of the graphs. Each student should commit to which graph(s) he/she will demonstrate prior to the presentations. You can then pick a random graph for each group. If time allows you may have each group demonstrate more than one graph.

Processing. Besides whatever comments naturally come out of reviewing all the reports, it seems important in the processing to focus on examples and non-examples of persistence in each group. For example, each group can identify one example of persisting and one non-example of persisting in their group work. A class discussion based on the examples can follow. The processing should include a discussion of the difference (and relationship) between the functions $D = 2^x$ and $D = \log_2 x$. In addition, you can give the following three questions to individuals to test for transfer and understanding: Describe how you would move to model $D = t^5$, $D = t^3$ and $D = 3t + 5$.

Suggested Format. The students work in groups of three and are given a portion of two or three periods to work with the probes and a portion of one period to prepare their demonstrations. The presentations and the processing will each require a portion of a class period. The project can be completed in six to ten school days.

Enrichment and Extension: Each group can work on additional functions such as $D = (2/3)t + 2$, $D = \sin t$, and/or $D = e^{-t} \sin t$.

It has been emphasized that content objectives do not play a significant role in the selection of nonroutine problems. However, this problem is an example of a good nonroutine problem that also is an excellent vehicle for teaching the concept of graphs. When students need to model functions through movement, they generally gain an excellent understanding of the functions.

Advanced

There are nine problems at this level of difficulty. The first three require students to explore functions given only their abstract definition (e.g., F(ab) = F(a)F(b) and a few values of F(x)). In addition to being excellent nonroutine problems, they provide excellent practice in thinking abstractly:

(1) **"F(x)":** Students are given the abstract definition of a function, F(ab) = F(a)F(b) and three values of the function, F(2), F(3) and F(5). They are asked to determine as best they can the values of the function for x = 1, 2, 3, ... and 20. Some values cannot be determined with just the given values; therefore, the students must develop techniques to approximate the values. The purpose of this task is to give the student practice solving a nonroutine problem involving abstract functions. This is an individual project but certainly can be completed in small groups if more appropriate for your context.

Sample student problem statement. The task for this problem is to determine some values for a function, given only the abstract definition and a few values of the function. Specifically: (a) You are given the following information about a function F(X): For all a, b, F(ab) = F(a) F(b), F(2) ≅ 2.585, F(3) ≅ 4.505 and F(5) ≅ 9.070. As best you can, complete the values of F(X) for X = 1 to 20, rounding your answer to the nearest 1000th. Some values you probably will not be able to determine to the nearest 1000th directly from the given information. Part of the assignment is to determine your best approximation for these values. (b) You will turn in a written report including the strategies you used and how you determined each of your values. (c) Only a few students will be asked to orally explain their project but each student should be prepared to give an oral presentation.

Teaching suggestions. Prior to starting the task, students need to have some understanding of functions and function notation. For example, typically this problem would be inappropriate for an algebra 1 class but appropriate at some point in an algebra 2 class. To insure that they have a basic understanding of the meaning of the abstract property of F(X), review how to calculate F(6): F(6) = F(3×2) = F(3) × F(2) ≅ 4.505 × 2.585 ≅ 11.645. Explain that some of the values (e.g., F(13), F(17),...) cannot be determined by relying only on the given values. Tell the students that the function is not linear and that for this assignment it is not acceptable to determine an unknown value by simply averaging known values. For example, it is unacceptable to calculate F(17) as (F(16) + F(18))/2. Remind students that part of the assignment is for them to document their work, including all non-trivial strategies and justification for the strategy selected. Answer any questions the students may have.

Comment. The function that I chose is given by $F(X) = X^{1.37}$. When this problem was first field-tested F(X) was defined as $F(X) = X^{1.5}$. With this definition a number of students were able to discover the function and thereby determine the exact rounded-off values as $F(X) = X^{1.5} = X\sqrt{X}$. Although this type of discovery certainly represents good mathematical thinking, it does not allow the students to address the meat of the problem from the viewpoint of nonroutine problems; that is, to develop a method for determining values of F(X) which cannot be determined directly. The choice of $F(X) = X^{1.37}$ minimizes the possibility that the students will determine the actual function. Even when a student does discover that $F(X) = X^{1.37}$, usually it is only after considerable effort trying other strategies. When the projects are complete, collect the written reports. If, in reviewing the reports, any oral presentations seem appropriate, it is probably best to schedule them before the general processing of the project.

Processing. Besides whatever comments naturally come out of reviewing all the reports, it seems important in the processing to focus on how students organized their data to discover a pattern for predicting the values of the function.

Suggested Format. The students work individually on this task. If it seems appropriate for your class, this problem can be assigned as a group project. The oral reports, if any, and the processing will require a portion of a class period. The project can be completed in five to ten school days.

Enrichment and Extension. One natural extension for this problem is a discussion of powers. For example, if presented correctly, students are fascinated with the question of why a number raised to the zero power is equal to one. You can start the discussion by asking the students why $5^0 = 1$, of course, not accepting "That's what we learned in algebra" as an answer. In general, students do not know why. You can then either demonstrate or lead the students to the fact that although the idea of raising something to the zero power does not make intuitive sense, we can see that inductively it makes sense by looking at the following pattern:

$5^4 = 625$
$5^3 = 125$
$5^2 = 25$
$5^1 = 5$
$5^0 = ?$

In other words, the definition is consistent with the pattern of exponents. Similarly, one can see that the definition is consistent with the law of exponents: $x^a / x^b = x^{a-b}$ (x not equal to 0) by noticing that $1 = x^a / x^a = x^{a-a} = x^0$. Equally interesting subjects for discussion include raising numbers to imaginary powers, irrational powers or transcendental powers such as π.

Notice that F(X) is a nonintegral rational root. An interesting historical note is that even what we consider commonplace numbers such as the square root of two were not discovered until the time of Pythagoras. Previously the common hypothesis was that all real numbers could be expressed as rational numbers. Pythagoras was able to show that the length of the hypotenuse of an isosceles right triangle with legs of one unit was not a rational number. He did this by means of an indirect proof. He assumed that the hypotenuse h was rational; i.e., $h = p/q$ or $h^2 = 2 = (p/q)^2$, where p and q are integers. Assuming that p/q is in reduced form, he argued that p and q could not both be even (if so, p/q could be reduced by dividing p and q by 2). The equation $2 = (p/q)^2$ can be rewritten: $2q^2 = p^2$. He then looked at the three remaining possibilities for p and q. The first possibility is that p and q are both odd; this is not possible because $2q^2$ would be even and p^2 would be odd. The second possibility is that p is odd and q is even. This is not possible because $2q^2$ would be even and p^2 would be odd. The third possibility is that p is even and q is odd; this is not possible because if p is even then $p = 2r$ for some r, so we can rewrite our equation to get $2q^2 = 4r^2$; dividing each side by 2 we get $q^2 = 2r^2$, a contradiction because q^2 is odd and $2r^2$ is even. By eliminating all four possibilities, we have contradicted the original assumption thereby proving that the square root of 2 is not rational.

A good extra problem for students can be created by you demonstrating the proof except for the last possibility or last two possibilities and have them try to complete the remainder of the proof.

Solutions. The following are straightforward calculations that all students should be able to determine: $F(4) = F(2)F(2)$, $F(6) = F(2)F(3)$, $F(8) = F(2)F(2)F(2)$, $F(9) = F(3)F(3)$, $F(10) = F(2)F(5)$, $F(12) = F(2)F(2)F(3)$, $F(15) = F(5)F(3)$, $F(16) = F(4)F(4)$, $F(18) = F(2)F(9)$, $F(20) = F(4)F(5)$. In addition, many students realize that $F(1)$ can be calculated by noticing that $F(a) = F(a1) = F(a)F(1)$; therefore, $F(1) = 1$ (assuming $F(X)$ is not equal to 0). The challenge of the problem is to determine good approximations for $F(7)$, $F(11)$, $F(13)$, $F(14)$, $F(17)$ and $F(19)$. Remember from the directions that the students are not allowed to use averaging of two given values; e.g., a student cannot calculate $F(17) = [F(16) + F(18)]/2$.

The most common strategy for determining the difficult values is to attempt to discover a pattern of how the values increase. For example, by graphing the given and computed values carefully on a Cartesian coordinate system one can sketch a smooth graph of $y = F(X)$ and thereby approximate the intermediate values. Or by studying the values of $F(X + 1)/F(X)$ and/or $F(X + 1) - F(X)$ for $X = 1$ to 20 (when possible), one can approximate some unknown values. Some students include values of $F(X)$ for fractional values of X such as $X = 3/2$; thereby increasing the accuracy of their approximations.

Two more sophisticated solutions are: (a) to notice that the graph is similar to that of a parabola and to use three given values to determine a parabola (or parabolas) that approximates the actual graph and (b) to use values of $X > 20$ to approximate unknown values; e.g., $F(121) = 2F(11)$ and $F(120) = F(10)F(12)$, so $F(11) = [F(10) F(12)]/2$. In addition, some students conjecture that $F(X) = X^n$, usually after first pursuing one or more other strategies, then use a calculator to determine the value of n. One strategy that students use that is not successful is that they become convinced that the graph of the function is a parabola (versus using parabolas which approximate the curve) and are obviously unsuccessful in finding a parabola that works. It should be mentioned that the better student might make this mistake but effectively use the error to realize that $F(X) = X^n$ and then determine the value of n.

Because the given values are approximate (rounded off to the nearest 1000th), there can be some discrepancy in the "best" approximations. For example, using the function $F(X) = X^{1.37}$, you get slightly different values then if you use the abstract property and the given values. The primary values given below were determined using $F(X) = X^{1.37}$; the values in parentheses were determined directly from the abstract definition and the given values.

F(1) = 1.000	F(11) = 26.712
F(2) = 2.585	F(12) = 30.094(30.102)
F(3) = 4.505	F(13) = 33.582
F(4) = 6.681(6.682)	F(14) = 37.170
F(5) = 9.070	F(15) = 40.855(40.860)
F(6) = 11.643(11.645)	F(16) = 44.632(44.649)
F(7) = 14.381	F(17) = 48.497
F(8) = 17.268(17.273)	F(18) = 52.447(52.461)
F(9) = 20.291(20.295)	F(19) = 56.480
F(10) = 23.442(23.446)	F(20) = 60.591(60.606)

Processing. One excellent focus for the discussion is the problem solving step of persistence. Students tend not to persist for the beginning problems, generally not being able to accurately judge the adequacy of their solutions. For example, some weaker solutions use graphing for their final solution. For a typical graph, the values obtained are between .1 and .2 off of the actual answers (compared to less than .02 for better solutions). In contrast, a student who eventually discovered the function, started with graphing:

I decided that making a graph with the points that I knew might give me an approximation or a picture of the function.

The accuracy of the graph was nowhere near the accuracy of the three decimal points but I wasn't expecting to get the values from the graph. By using the graph, I got an approximation of some of the missing values.

This student's strategy is a good example of the problem solving steps. The student approaches this difficult problem by trying something, in this case graphing. The student is aware that the problem has not yet been completely solved, graphing is not an accurate solution, and the student persists until eventually discovering the actual function.

Another student's remarks illustrate well the feeling that many times accompanies the step of persistence:

However a great flaw appeared after I had determined F(7)… Subtracting from F(8) the continuous pattern no longer continued. My approach was obviously off course. Now this was becoming frustrating for deep down I knew there was a simple solution though I had yet to find it.

Notice that unlike the typical student who might ignore a discrepancy, this student recognizes that there is a problem. By the way, the student persisted and discovered the function.

From the point of view of the algebra curriculum, this problem gives the students good experience working "concretely" (determining specific values) with a difficult abstract definition of a function. In addition, by the end of the processing of this problem, the students have a fairly clear understanding of the properties of a power function.

Another good focus for the discussion is the strategy of looking for patterns, such as in the increases in the values that can be determined directly. One student attempted to organize the data to find patterns. There was an obvious discrepancy in the last column, the numerical value increase for F(13) and F(14) are less than some previous increases and therefore do not fit the pattern. In the margin next to these values the student wrote: "the increase is stopped here, can't understand." This solution is a good example of a student trying something (the chart) but not persisting until reaching an adequate solution. Despite the excellent effort required to create the chart, the student did not take the needed last step, to experiment with adjusting some of the values so that the entire chart was consistent. In contrast, a second student organized the data similarly and initially ran into similar discrepancies; however, this student kept working with the data until he came up with a consistent solution. For example, for F(11) the first student determined F(11) = 26.646 and the second student determined F(11) = 26.711 compared to the actual value of 26.712. This example illustrates the relevance of persistence in inductive problems that involve complicated data.

F(X) Worksheet

I. You are given the following information about a function F(x):
    (a) For all a, b > 0, F(ab) = F(a)F(b)
    (b) Three values for F(X), F(2) = 2.585, F(3) = 4.505 and
    F(5) = 9.070.
    Note: Above "=" means approximate value, rounded off to the
nearest 1000th.

As best you can, complete the following values.

F(1) =   F(11) =
F(2) = 2.585 F(12) =
F(3) = 4.505 F(13) =
F(4) =   F(14) =
F(5) = 9.070 F(15) =
F(6) =   F(16) =
F(7) =   F(17) =
F(8) =   F(18) =
F(9) =   F(19) =
F(10) =   F(20) =

(2) **"G(x)"**: Given that G(ab) = G(a) + G(b) and G(8) ≅ 1.660, G(14) ≅ 2.107 and G(6) ≅ 1.430, as best you can, complete the values of G(X) for X = 1 to 20, rounding your answer to the nearest 1000th. Averaging is an unacceptable method.

Sample student problem statement. The task for this problem is to determine some values for a function, given only the abstract definition and a few values of the function. The purpose of this task is to give you practice solving a nonroutine problem involving an abstract function and careful calculations. Specifically:

1 You are given the following information about a function G(X):
    For all a, b, G(ab) = G(a) + G(b), G(8) ≅ 1.660, G(14) ≅ 2.107
    and G(6) ≅ 1.430. As best you can, complete the values of G(X)

for X = 1 to 20, rounding your answer to the nearest 1000th. Some values you probably will not be able to determine to the nearest 1000th directly from the given information. Part of the assignment is to determine your best approximation for these values.

2   You will turn in a written report including the strategies you used and how you determined each of your values.

3   Only a few students will be asked to orally explain their project, but each student should be prepared to give an oral presentation.

Teaching suggestions. This assignment is similar to F(X) but is more difficult because G(X) is a log function which most students have more difficulty understanding and the given values for G(X) are more difficult to work with than the given values for F(X). In most cases, it is appropriate to have the students complete "F(X)" prior to this problem. Give the handout G(X) Worksheet to the students and briefly explain the problem. For F(X) the students were shown how to calculate F(6) to insure that the students had enough of an understanding of the definition of F(X) to successfully complete the straightforward part of the assignment. It is suggested not to give the students any direct help for this problem, given their experience with F(X). You might clarify the assignment by giving them a made up example; e.g., if G(2) = 2.310 and G(3) = 3.257, then G(6) = G(3) + G(2) = 2.310 + 3.257 = 5.567. The rationale for this suggestion is that a key component for this problem is to determine how to use the given values to obtain the majority of the values. For example, the students need to see that G(8) = 3G(2); therefore, G(2) can be calculated to be G(8)/3.

Comment: It may seem as if this problem is fairly easy considering that the students have completed F(X). However, except for the exceptional student, this problem is at a very appropriate level of difficulty. The average student either never realizes or does not immediately realize that the function is a log function and thereby quickly solve the problem. On the other hand, experience with F(X) helps the student see possible strategies to solve the problem and eventually reach an adequate solution. Consequently, this problem is not only challenging, but also helps improve the student's confidence in his or her ability to solve nonroutine problems. Also, the nonroutine problem "Marking G(X)," which is described next, involves the evaluation of solutions to this G(X) problem. You might want to read the directions for "Marking G(X)" before returning students' work for this problem.

When the projects are complete, have the groups give their oral presentations. When the presentations are complete, collect the written reports.

Suggested format. The students work in groups of three and are given a portion of two periods to work on the project and prepare their oral report. The oral reports and the processing will each require a portion of a class period. The project can be completed in five to ten school days. Of course, if it seems appropriate for your class, this nonroutine problem can be assigned as an individual project.

Solutions: The key to starting this problem is to realize that $G(8) = G(2\times2\times2) = 3G(2)$, so $G(2) = G(8)/3$. Then $G(3)$ can be calculated by realizing that $G(6) = G(3) + G(2)$, so $G(3) = G(6) - G(2)$. Similarly, $G(7)$ can be calculated. Then by direct substitution one can calculate $G(4)$, $G(9)$, $G(12)$, $G(14)$, $G(16)$ and $G(18)$. Also, it can be seen that $G(1) = 0$ (e.g., $G(a1) = G(a) + G(1)$; so $G(1) = 0$). As in the problem $F(X)$, the remaining values can be determined by trying to discover a pattern in how the function increases. Of course, if the student realizes that $G(X)$ must be a log function ($G(X) = \log_b X$), then it becomes fairly straightforward to determine the remaining values. For example, most students who determine that $G(X)$ is related to logarithms then decide to work with logarithms base 10, specifically they determine that $G(X) = (1.838)\log(X)$, obtaining: $G(1) = 0$, $G(2) = .553$, $G(3) = .877$, $G(4) = 1.107$, $G(5) = 1.285$, $G(6) = 1.430$, $G(7) = 1.553$, $G(8) = 1.660$, $G(9) = 1.754$, $G(10) = 1.838$, $G(11) = 1.914$, $G(12) = 1.984$, $G(13) = 2.047$, $G(14) = 2.107$, $G(15) = 2.162$, $G(16) = 2.213$, $G(17) = 2.262$, $G(18) = 2.307$, $G(19) = 2.350$ and $G(20) = 2.391$.

Processing. Besides whatever comments naturally come out of reviewing all the reports, it seems important to help the students understand the three steps of problem solving and the other problem solving skills, especially if this can be done by drawing examples from their solutions. Some methods that might be particularly useful for this problem include: (a) Compare and contrast the student solutions to this problem with their solutions to $F(X)$. Did the students demonstrate more persistence? (b) Focus on the manipulative skills by discussing whether typical solutions (e.g., graphing, studying patterns of increase) could have been improved in accuracy or contrast two solutions that used similar methods but one was more accurate. (c) Discuss solutions in which students discovered that $G(X)$ was a logarithm especially if the student attempted other solutions first, focusing on what preceded the solution. For example, many students realize that the function is logarithmic as a result of trying some other strategy which is not successful but helps the student realize a more productive strategy.

It can be useful to compare solutions that reach similar conclusions but by different steps. For example, here is an outline of three solutions which all conclude that $G(X)$ is equal to some constant times the log of X. In the first solution, the student very quickly "knew this problem could be solved using logarithms, but I wanted to first get approximations for some values of the function that I could use to test against the function I developed." After solving for the straightforward values, the student solved for k in $k\log(X) = G(X)$ by using $X = 6$ ($k = G(6)/\log6 = 1.838$) and checked the answer for other values of X. Only the exceptional student solves this problem so directly. The solution of the second student is more typical. First the student did not see how to solve for $G(2)$ so first tried generating equations involving unknown values of $G(X)$ but always ended up with more unknowns than equations. Then the student worked with fractional values, for example, $G(4/3) = G(8/6) = G(8) - G(6) = .230$. These values were used to generate a fairly good graph of $G(X)$. Then came a realization: $G(27) = G(3/4) + G(36)$ and $G(27) = 3G(3)$, thus $G(3) = .877$, after which the student realized how to figure out $G(2) = G(8)/3$ and the other straightforward values. At this point the student looked at the graph of $G(X)$ and came up with the idea that $G(X)$ was related to log X and soon discovered that $G(X) = 1.838\log(X)$. Notice that while this solution is certainly not elegant or straightforward, it does well illustrate a student using the three steps of problem solving, a significant accomplishment. The third solution is similar to the second solution in that the student attempted a number of strategies before discovering the successful strategy; however, the difference is that the third student obtained a value of 1.84 versus 1.838. The third student did not persist to the same extent as the second student and; therefore, did not reach as accurate a solution. This example is a good opportunity to discuss the importance of persistence, the third step in problem solving.

Other strategies that students have used include: (a) Noticing that the difference between consecutive integral values of $G(X)$ decreases as the value of X increases and using this fact to estimate the value of $G(X)$. For example, $G(7)$ can be estimated by calculating $G(49) = (G(48) + G(50))/2$ and noticing that $G(49) = 2G(7)$. (b) Some students very carefully graph $G(X)$ getting within .005 of the actual values. (c) One student used simultaneous equations. Although you cannot "solve" the equations (more unknowns than equations), you can obtain a very good solution by trial and error.

Enrichment and Extension. Obviously, this problem is an excellent introduction for the topic of logarithms and would be appropriate to assign just prior to such a unit. Even if logarithms are not part of the written curriculum for the course, it seems appropriate to take at least some time to discuss the topic. I like to draw the students' attention to the relatively slow growth of logarithms compared to other increasing functions such as $y = x^2$ and $y = 2^x$. If you are familiar with binary sorts in computer programming, you realize that the time for a binary sort grows like the log base two; therefore, for large sorts the binary sort takes substantially less time to complete when compared to sorts that grow like other functions.

John Napier (1550–1617) discovered logarithms and this discovery was considered one of greatest discoveries the world had seen. David Burton (1985, pp. 326–7) writes: "Seldom has a new discovery won such universal acclaim and acceptance. With logarithms, the operations of multiplication and division can be reduced to addition and subtraction, thereby saving an immense amount of calculation, especially when large numbers are involved. Since astronomy was notorious for the time-consuming computations it imposed, the French mathematician Pierre deLaplace was later to assert that the invention of logarithms "by shortening the labors, doubled the life of the astronomer." To give the students a better sense of the importance of the discovery, you could have the students do a relevant calculation without logs (and no calculators!) and then with logs.

G(X) Worksheet

I. You are given the following information about a function G(X):
    (a) For all a, b>0 G(ab) = G(a) + G(b)
    (b) Three values of G(X), G(8) = 1.660, G(14) = 2.107 and
        G(6) = 1.430.
    Note: Above "=" means approximate value, rounded off to the nearest 1000th.

    As best you can, complete the following values.
    G(1) =   G(11) =
    G(2) =   G(12) =
    G(3) =   G(13) =
    G(4) =   G(14) = 2.107

G(5) =   G(15) =
G(6) = 1.430 G(16) =
G(7) =   G(17) =
G(8) = 1.660 G(18) =
G(9) =   G(19) =
G(10) =   G(20) =

II. Explain (on a separate sheet) how you determined each of your
    values.

(3) **"Marking G(X)."** The task for this problem is to develop a short yet
    valid method for grading the G(X) assignment.

    Sample student problem statement: The purpose of this task is to
    give you practice solving a nonroutine problem involving functions
    and developing procedures. Specifically:

    1   Your group will develop a short yet valid method for grading the
        G(X) assignment. Your work will be evaluated on two criteria.
        First, "Is the method valid?" That is, would the results of your
        method be similar to the results of a panel of experts? Second,
        "How long does your method take to grade a group of papers?"
        The less time required to grade the papers the better.

    2   The information you were given for the G(X) problem and the
        actual values (rounded off values) of G(X) are given below:
        For a, b; G(ab) = G(a) + G(b), G(8) ≅ 1.660, G(14) ≅ 2.107 and
            G(6) ≅ 1.430.

        G(1) = 0   G(11) = 1.914
        G(2) = .553   G(12) = 1.984
        G(3) = .877   G(13) = 2.047
        G(4) = 1.107   G(14) = 2.107
        G(5) = 1.285   G(15) = 2.162
        G(6) = 1.430   G(16) = 2.213
        G(7) = 1.553   G(17) = 2.262
        G(8) = 1.660   G(18) = 2.307
        G(9) = 1.754   G(19) = 2.350
        G(10) = 1.838 G(20) = 2.391

    3   Each group will prepare an oral presentation of two minutes
        or less on the gathered information. I will select the member of
        your group to present the oral report.

    4   In addition, your group will turn in a written report including
        how you developed your final procedure and other methods you
        considered.

    Teaching suggestions. One way to introduce the problem is as
follows: Tell the students that this nonroutine problem will be based

on the results of the problem involving G(X), particularly how to mark the papers. Discuss the fact that theoretically an excellent way to mark the papers would be to have a panel of experts read all the papers and then score each paper. Point out that this method would be impractical timewise if you had, for example, 100 papers to mark. Suggest the following alternative: for each student calculate the sum for X = 1 to 20 of S = |[actual value of G(X)] − [student's value of G(X)]|, then give grades according to the value of S − the lower the value of S, the higher the grade. Through further discussion, students should realize that this method of marking would be reasonably accurate and certainly quicker than using a panel of experts; however, this method would still be very lengthy for marking 100 papers.

Remind the students that they will be graded on the validity of their method of marking and the length of time required to grade a paper using their method (the shorter, the better). Your assessment of your students should determine how much information you provide students to help them start the problem. Generally, two types of information can be useful. First, information concerning the G(X) problem such as: (a) the values of G(X) for X equaling 2, 3, 4, 6, 7, 8, 9, 12, 14, 16 and 18 can be determined directly and fairly easily from the given values, (b) G(1) can be determined from the given information but is not as straightforward a calculation, (c) the values of G(X) for X equaling 5, 10, 11, 13, 15, 17, 19 and 20 are more difficult to determine. in addition, 10 and 11, and 19 and 20 are adjacent difficult values; and the value for X equals 5 effects three other values (10, 15 and 20) and (d) the larger the value of X, the more accurate averaging becomes for determining unknown values; e.g., G(5) = 1.285 while [G(4) + G(6)]/2 = 1.269, whereas G(19) = 2.350 while [G(18) + G(20)]/2 = 2.349, only.001 off.

Also, you might want to give them some sample student solutions as data. One option would be to give the students copies of the solutions of their class to the G(X) problem. That way the data is relevant and more real. If you pick this last option, you might want to use your grades as a method of comparison (you being the panel of experts!) and even consider withholding your grades for the G(X) problem until after they have completed this assignment.

When the projects are complete, have the groups give their oral presentations. When the presentations are complete, collect the written reports.

Solutions. Most student solutions revolve about reducing the number of calculations by looking at just a few keys values. For example, you could take: (a) one value from G(2), G(3), or G(7); (b) the value of G(1); (c) one value from G(4), G(6), G(8), G(9), G(12), G(14), G(16) or G(18) and (d) perhaps two values from G(5), G(10), G(11), G(13), G(15), G(17), G(19) or G(20). Calculating the total sum, S = |G(X) − student's value of G(X)| for these five numbers would be an example of one possible solution. For example, one student

> devised a method so only eight additions and one subtraction is necessary. But the accuracy shouldn't be affected…. Check the eight key numbers which the student must calculate. From these eight numbers everything else can easily be found. The numbers are G(2), G(3), G(5), G(7), G(11), G(13), G(17), and G(19). Calculate the correct sum of these eight values and subtract the sum of the student's eight numbers.

Another alternative is to pick a similar set of numbers but to assign different weights to some values. For example, if G(20) was considered the most significant number then |G(20) − the student's G(20)| could be multiplied by three to increase its effect on the grade. One student tripled the weight of G(11) because "it is the lowest prime value that cannot be determined [directly] that is next to another value which cannot be determined [G(10)]."

Some students develop methods that can be carried out by inspection of a few key values. For example, one student developed a system based on the values of G(1), G(5) and G(16), reasoning that

> If a student determines that G(1) is 0, he thoroughly understands the relationship G(ab) = G(a) + G(b)….If you determine G(16) to be 2.213 you have had to figure out that the function is log base 3.5, because when you determine the values arithmetically… you find that G(16) = 2.214….Averaging G(14) and G(16) gives a value for G(15) that is only off by.001. But when you average G(4) and G(6) the value for G(5) is off by.016.

Based on these observations the student developed the following grading system with a 4.0 scale: (a) 1 point if G(1) = 0, (b) 1/2 point if G(5) = 1.285 and G(16) = 2.213, (c) 1 possible point: 1 point if G(16) is between 2.212 and 2.214 or 1/2 point if G(16) is between 2.208 and 2.218 and (d) 1 1/2 possible points: 1 1/2 points if G(5) is between 1.275 and 1.295, 1 point if G(5) is between 1.270 and 1.300, 1/2 point if G(5) is between 1.269 and 1.301 and 1/4 point if G(5) is between 1.260 and 1.310.

Processing. One suggestion for the discussion is to have the students try out their methods on actual or made up solutions and compare the results to the teacher's assessment or a longer method of calculation (see the enrichment section). It could be beneficial for students to try out other student's methods either in pairs or small groups. One focus could be the clarity of the methods, an important skill in developing algorithms.

Suggested Format. The students work in groups of three and are given a portion of two periods to work on the project and pre-pare their oral report. The oral reports and the processing will each require a portion of a class period. The project can be completed in five to ten school days. If appropriate for your context, this can be assigned as an individual problem.

Enrichment and Extension. Each group can figure out a similar marking system for F(X) and the next problem, R(X).

(4) **"R(X)":** Students are given the abstract definition of a function, $R(X) = R(X - 1) + X$ and one value of the function and asked to deter-mine a variety of difficult values (e.g., R(1000)).

Sample student problem statement. The task for this problem is to determine some values of a function R(X), given only the abstract definition and one value. The purpose of this task is to give you practice solving a nonroutine problem involving an abstract func-tion. Specifically:

1   Your group is given the following information about the function R(X): $R(X) = R(X - 1) + X$, and $R(1) = 2$. Determine the values of R(X) for X = 1 to 20.

2   The following values of R(X) are more difficult to determine. Some require you to discover a pattern. Some cannot be calcu-lated with the given information and require you to extend the definition of R(X). Determine the best values you can for the five values of the function below. If you extend the definition of R(X), you are required to give a justification for the extension. Also, document any pattern you use. The five values you need to determine are: R(−1000), R(2000), R(55 1/2), R(7 4/5) and R(8.732).

3   Each group will prepare an oral presentation of two minutes or less on the gathered information. I will select the member of your group to present the oral report.

4   In addition, your group will turn in a written report including how you arrived at your calculations and your justification for those calculations.

Teaching suggestions. Give the students the problem statement for R(X). Go over the directions and answer any student questions. Explain that the first part of the assignment, calculating R(X) for X = 1 to 20, is straightforward. You might calculate R(2), R(3) and R(4) as a class to be certain that they understand the recursive definition. Point out that the difficult part of the problem is the additional values and that they can use the data from the first part to help determine the difficult values. Tell students that it is not recommended that they figure out a value such as R(2000) by determining all the values from R(1) to R(2000). You might add that a good strategy is to look for patterns. Let the students know that the function is not linear and; therefore, the student cannot simply calculate, for example, R(7 4/5) as R(7) + 4/5[R(8) − R(7)]. You might point out to the students that R(X) is a discrete function that is not defined for non-integral values. Therefore, when they assign a value, for example, to R(55 1/2), they need to justify this extension of the function to a nonintegral value.

When the projects are complete, have the groups give their oral presentations. When the presentations are complete, collect the written reports.

Solutions: It turns out that $R(X) = .5X^2 + .5X + 1 = X(X + 1)/2 + 1$ for positive integral values of X. Students that discover this pattern do so in a variety of ways. One student drew a graph with easily obtained values, hypothesized that the graph was a parabola, and solved for a, b and c in $R(X) = aX^2 + bX + c$ by substituting three pairs of values. Most students who determine R(X) do not find the solution so directly. For example, one student first experimented with the two equations:

$$R(X) = R(X − 1) + X$$

$$R(X + 1) = R(X) + X + 1$$

to obtain:

$$R(X + 1) − R(X) = R(X) − R(X − 1) + 1$$

but the student concluded: "This makes perfect sense since each time X increases by 1, it increases the difference between two consecutive values by 1 too. But it doesn't help much." Then the student tried graphing and hypothesized it was a parabola, but first worked with the forms $y = x^2 + c$ and $y = x^2 + x + c$ before trying $y = .5x^2 + .5x + c$. This second student's solution is more typical of the "meandering" involved in most student solutions.

Other students discover the pattern that $R(X)$ is one more than the sum of 1, 2, 3, ... and X; that is, $R(X) = X(X + 1)/2 + 1$. For example, one student graphed the easy values, thought it was a parabola but could not figure out the equation. "Then [the student] looked at what $R(X)$ equaled.

$$R(5) = 16 = R(4) + 5 = R(3) + 4 + 5 = R(2) + 3 + 4 + 5$$

well it turned out that $1 + 2 + 3 + 4 + 5 + 1$ equaled $R(5)$ and it follows that $R(X)$ equals the summation of numbers from 1 to X, plus one more." After more work the student discovered that $1 + 2 + 3 +... + X = X(X + 1)/2$ and completed the assignment.

The most common solution by students who do not discover the equation is by graphing. Of course, the accuracy of this solution depends upon the manipulative skills of the student. How carefully are the points graphed? What scale is used? etc. For example, one student obtained a value of 35.3 (versus 35.32) for $R(7\ 4/5)$ by careful graphing, including a check by determining $R(1\ 4/5)$ by graphing and using the definition to get $R(7\ 4/5)$.

Processing. Besides whatever comments naturally come out of reviewing all the reports, discuss different methods students used to calculate the values of $R(X)$. A discussion of the difference between a discrete function and a continuous function seems appropriate. It should be noted that $R(X)$ is a discrete function not defined for nonintegral values. For example, if you wrote a PASCAL program with the function $R(X)$, you would not be able to get a value for $R(7\ 4/5)$. Since it has the same value as $X(X + 1)/2 + 1$ for positive integers, it makes sense to extend the discovered value of $R(X)$ for positive integers to nonintegral values.

Suggested Format. The students work in groups of three and are given a portion of two periods to complete the project and prepare their oral report. The oral reports and the processing will each require a portion of a class period. The project can be completed in five to ten school days. This problem can be completed individually if students have appropriate previous experience solving nonroutine problems, such as $F(X)$ and $G(X)$.

Enrichment and Extension. One extension of this problem is to study recursive functions in a computer language such as PASCAL.

One activity suggested in the discussion is to write a program which prints out values of R(X) and to see which values can or cannot be printed out (nonintegral cannot). This activity could lead to a study of discrete functions defined only for positive integral values. The Fibonacci sequence defined by F(N + 1) = F(N) + F(N −1) with F(1) = 1 and F(2) = 1 provides an excellent topic for enrichment.

(5) **"Area Under a Curve":** This nonroutine problem requires the students to calculate the area under the curve $y = x^2 + 2$ between $y = 0$, $x = 0$ and $x = 3$.

This problem is a key nonroutine problem that I have already discussed in Chapters 1 and 2. My initial experience implementing this problem with my former calculus students is the primary experience that indicated to me the potential significance of, and need for, a curriculum of nonroutine problems and motivated me to devote so much time to developing that curriculum. Specifically, this problem illustrates that even the best secondary mathematics students have difficulty (without instruction) with the step of persistence. For example, despite directions to determine the area "as best they can," most calculus students calculate the area as 15 1/2 square units by "replacing" the curve with the straight lines connecting (0,2) and (1,3), (1,3) and (2,6), and (2,6) and (3,11) and then dividing the area into triangles and rectangles (or trapezoids). An "obvious" better solution would be to divide the area into a larger number of trapezoids (e.g., six divisions yield a solution of 15.125). It should be noted that most students divide the area into three trapezoids (or rectangles and triangles) because they do not have a sense that their solution is not adequate, versus unwillingness to make the necessary effort. In contrast, the step of persistence implies having a connection with what a satisfactory solution looks like. These students fail to recognize that their calculations could be more accurate by simply increasing the number of divisions. Notice their solution is inadequate even though the strategy they picked is excellent and, in fact, is the basis of the method of integration in calculus that gives an exact answer to the problem. From having worked with these students, I believe most of the students, typically high achievers, would have been willing to spend the time necessary to persist but simply did not have experience solving problems which require them to recognize that a problem does not always have a straightforward solution. Certainly typical textbook problems do not require this skill. Consequently, in processing this portion of the problem, it is important to help the students realize that the strategy of approximating the area by using trapezoids is an excellent strategy but that the strategy needs to be carried out until an adequate solution is reached; that is, the step of persistence is needed to arrive at a good solution.

In processing this problem, the teacher might first outline the typical solution of 15 1/2 square units. Then, if any students used this strategy with more than three trapezoids, their solutions could be discussed, emphasizing the increased accuracy and any stated student realization of the need to increase the number of trapezoids. Here are some examples of quotes from student solutions that would be appropriate:

> "This solution [15 1/2 square units] seemed like it could be improved. When I blew up the graph I could see that I was over on my answer…. I decided to divide the graph by thirds" (i.e., nine trapezoids).
> "At a certain point I saw that the area was approaching 15 [this student wrote a computer program that increased the number of divisions]. I would increase the number of divisions and get basically the same answer [e.g., 14.99]….So my answer is 15."
> "But I wasn't satisfied [with the solution of 15 1/2]. So I pressed on by trying more divisions…"

Also, in the processing, if a student with more than three divisions did not state the reason for more divisions in the documentation, the teacher could question the student, trying to solicit the student's reason. The emphasis in this portion of the processing is to make the students aware of the affective component necessary for an adequate solution, dissatisfaction with the initial solution. Ideally, sharing examples of student documentation or questioning of students can give the class a sense of this component. If not, the teacher needs to model the affective component by illustrating how a mathematician using this strategy might have proceeded.

In practice, approximately 75% of my calculus students determine the best approximation of the area to be 15.5 square units when I give it as the first nonroutine problem that they attempt. Of the remaining students, a few even determine the area to be 15 square units (actual area). I will note that when I gave this problem as part of a summative evaluation to students I taught in an undergraduate problem solving course for non-mathematics majors that included a curriculum of nonroutine problems a higher percentage determined an area better than 15.5.

(6) **"Slope":** Students develop a method to determine the slope of a variety of nonlinear functions.

Sample student problem statement. The task for this problem is to develop a procedure for estimating the slope of some quadratic equations. The purpose of this task is to give you practice solving a nonroutine problem involving functions and geometry. Specifically:

1 Your group will develop a written procedure for estimating the slope of the following functions: $y = x^2$ for any integral value of x less than 50 and $y = ax^2 + bx + c$ for which a, b, c and x are integers and a, b, and c can take the values 1 to 10; and x can take the values 1 to 5.

2 Your procedure must be usable by another person (e.g., another student) and will be tested by having the procedure tried by another student with values of a, b, c, and x that I provide.

3 Each group will prepare an oral presentation of two minutes or less on the final method you used to calculate your answers. I will select the member of your group to present the oral report.

4 Your group will turn in a written report including a description of your final method, why you selected that method and a description of any other methods you considered.

Teaching suggestions. Prior to starting the task, students need to understand, by visual inspection of a linear graph, whether the slope is 0, positive or negative and which graph of two linear graphs with positive (or negative) slope has the largest slope. If necessary, these concepts should be reviewed. Then demonstrate how one can use the concept of slope for linear equations to determine whether the slope of $y = x^2$ (e.g., draw the graph on the chalkboard) at a specific point is 0, positive or negative and which of two points on the graph with positive (or negative) slope has the slope with the largest absolute value. You can indicate that the slope of a nonlinear graph can be approximated by the slope of the tangent to the graph at the desired point, perhaps without using the term tangent but rather a drawing showing the tangent approximately. For example., students can see that the slope is more at larger positive values of x such as comparing the graph and tangents of $y = x^2$ at $x = 1$ and $x = 5$.

Go over the directions and answer any student questions. Be sure the students understand that their procedure will be tested by another student so it needs to be clear. Tell them that you will provide the values for the test. This is a good project to require a progress report.

When the projects are complete, have the groups give their oral presentations and collect the written reports.

Processing. Besides whatever comments naturally come out of reviewing all the reports, discuss the steps groups took to determine their final procedure. It would be valuable to share the results of a group that tried a logical sequence of methods that eventually led to a good procedure.

Suggested Format. The students work in groups of three and are given a portion of three periods to work on the project and prepare their oral report. The oral reports and the processing will each require a portion of a class period. The project can be completed in ten to fifteen school days.

Enrichment and Extension. Expanding the project to additional functions such as cubics could be interesting. Also, note that this problem would be a good introduction to the concept of slope in calculus.

(7) **"Exploring Functions":** The task for this problem is to explore and compare the following functions: $y = x$, $y = x^2$, $y = 2^x$, $y = x!$, $y = \log_2 x$, and $y = 2x$. The students prepare a description of the functions including a 100 word or less description of each function, an additional 200 word or less description of the interrelationships among the functions and a less than 500 word discussion of the more subtle and interesting characteristics of the functions. The purpose of this task is to give the students practice solving a nonroutine problem involving functions and expressing mathematical ideas concisely. This is an individual project.

Sample student problem statement: The task for this problem is to explore and compare the following functions: $y = x$, $y = x^2$, $y = 2^x$, $y = x!$, $y = \log_2 x$, and $y = 2x$. Specifically:

1   You are hired to write a two part description and comparison for students of the following functions: $y = x$, $y = x^2$, $y = 2^x$, $y = x!$, $y = \log_2 x$, and $y = 2x$.

2   In the first part, you will give a general description of the functions consisting of a 100 word or less description of each function and an additional 200 word or less description of the interrelationships among the functions. The first part should be accurate, concise and convey a clear general picture of each function and its major characteristics.

3   In the second part you will prepare a 500 word or less discussion of the more subtle and interesting characteristics of the functions. The second part should be interesting and give the reader a clearer, deeper understanding of the functions.

4   I will act as an editor for the project; that is, if you have a format different than the suggested format that you think would be effective, you can ask for my approval for the changes.

5   Only a few students will be asked to orally explain their project but each student should be prepared to give an oral presentation.

Teaching suggestions. I have used this problem as part of an assessment to measure whether students have successfully mastered the skills involved in solving nonroutine problems. The problem requires the student to determine what data to generate, organize the data and determine patterns or properties, evaluate the importance of the data and synthesize the data concisely and meaningfully. At this point in the curriculum, the student is simply given the statement of the problem and expected to complete the project without additional assistance. Most of my students who completed this project successfully demonstrated an excellent understanding of these key functions and the relationships among them.

Prerequisite skills. The students need some understanding of the functions. Specifically, the students should have studied all the functions, except perhaps $y = x!$. It should be sufficient for this project to give them the definition of $y = x!$ [$y = x(x-1)(x-2) \ldots (2)(1)$]. Go over the directions and answer any student questions. You might want to emphasize that they have the option of approaching you concerning the format of the project. You can let them know that this corresponds to the process many authors go through in writing a book in which a general format for the book is approved by the editor and, if the writer wants a change, he or she must convince the editor.

This project can act as a measure of the curriculum and an evaluation of the students' understanding of functions; therefore, you might want to require a progress report. When the projects are complete, collect the written reports. If, in reviewing the reports, any oral presentations seem appropriate, it is probably best to schedule them before the general processing of the project.

Processing. Besides whatever comments naturally come out of reviewing all the reports, it seems important in the processing to focus on the interesting discoveries students made and how they were able to discover them.

Suggested Format. The students work individually on this task. The oral reports, if any, and the processing will require a portion of a class period. The project can be completed in ten to fifteen school days.

Enrichment and Extension. A good extension would be to develop, as a class, materials actually to be used for a mathematics class in the future.

(6) **"y = sin x – e^x.": "Y = sin x – e^x"** is an example of a nonroutine problem that is appropriate for more advanced students with experience solving nonroutine problems. In fact, this nonroutine problem as well as "Exploring functions" are two problems I use to assess student's understanding of how to solve nonroutine problems, particularly for above average students in more advanced mathematics courses. The task is to find the 50th root less than 0 for the equation $y = \sin x - e^x$ (e is approximately 2.714 and is an important number in calculus). For the reader not familiar with this type of equation, you cannot solve the equation for x through normal algebraic manipulation, rather the student needs to generate data using a graphing calculator and look for patterns. In working with calculus students with experience solving nonroutine problems, a typical student solution is as follows: (a) Problem recognition: the student realizes the problem is not solvable by algebraic manipulation, sees the need to generate data and has some confidence that trying something and persisting will lead to a solution; (b) trying something: the student uses a graphing calculator to generate the first few roots and looks for patterns, noticing a fairly linear decrease in the value of the roots and (c) persistence: in trying to make sense of the pattern the student realizes that the roots change by a value approaching pi or perhaps approaching 3 initially, and realizes that e^x contributes close to zero to the value of sin x – e^x as x decreases in value (e.g., for x = –20, e^x is approximately.000000002); therefore, the appropriate root of y = –sin x (i.e., x = –50pi, which is easy to determine) is an excellent approximation of the actual root. In fact, a better approximation for the root is not possible even with a graphing calculator. When I have tried this problem with Calculus students with experience solving nonroutine problems, approximately 75% of the students found an excellent solution to this problem. My experience indicates that without instruction in solving nonroutine problems (or another approach with similar goals) the majority of students would not be able to solve this problem. My hypothesis is that an average academic student in the fourth year of a mathematics program that integrated nonroutine problems would be able to successfully solve a nonroutine problem of this difficulty level. In summary, this is an example of a nonroutine problem that requires a good understanding of content (to be able to work with a function such as $y = \sin x - e^x$)

and requires technology as a tool in the problem solving process. This problem is most appropriate later in a curriculum of nonroutine problems not only due to the difficulty of the content, but also because the problem is more powerful as a level six problem (see Chapter 2); that is, a problem with minimal scaffolding.

In addition, a good solution requires the student to use the calculator as a problem solving tool to help identify a pattern. In other words, this is an example of a nonroutine problem that naturally requires technology but also requires students to use higher order thinking skills to reach an excellent solution.

Sample student problem statement. The task for this problem is to find a specific root for the equation $y = \sin x - e^x$. The purpose of this task is to give you practice solving a nonroutine problem involving finding solutions to a complicated equation and finding a pattern. Specifically:

1 Your task is to find the 50th root to the left of $x = 0$ for the equation $y = \sin x - e^x$.

2 You will turn in a written report including a description of the process you used to solve this problem (including unsuccessful steps that you took), your final solution and why you believe your solution is good.

3 Only a few students will be asked to orally explain their project, but each student should be prepared to give an oral presentation.

Teaching suggestions. The students need to be familiar with using a graphing calculator and have a sense of the graphs of $y = \sin x$ and $y = e^x$. Go over the directions and answer any student questions. Depending on your students' experience, you might want to emphasize that it is impractical to calculate this root by using just the trace function and that they need to generate some data and look for a pattern.

Since e^x approaches zero quickly when x is negative the solutions for x less than 0 to $y = \sin x$ closely approximates the actual roots. The nth root less than $x = 0$ for $y = \sin x$ is $-n\pi$; therefore, $x = -50\pi$ is an excellent approximation for the 50th root. For example, some students generate the first few roots (the first three roots are approximately −3.1830, −6.2813 and −9.4249) then plot points of the form: (the number of the root, the value of the root) and fit a straight line to the points and notice that the slope is approximately π or perhaps initially a slope of 3. The value $x = 50\pi$ gives a value for y of $5.1579000625422 \times 10{-28}$ on the TI82 (with some manipulation). Using a graphing calculator without understanding the pattern and function it is not possible to find a value of x that will yield a better value for y for the 50th root.

When the projects are complete, collect the written reports. If in reviewing the reports any oral presentations seem appropriate, it is probably best to schedule them before the general processing of the project.

Processing. Besides whatever comments naturally come out of reviewing all the reports, it seems important in the processing to discuss how students generated and organized data to find a pattern and any good examples of persisting.

Suggested Format. The students work individually on this task. The oral reports, if any, and the processing will require a portion of a class period. The project can be completed in five to ten school days.

References

Burton, D. (1985). *The history of mathematics: An introduction*. Boston, MA: Allyn and Bacon.

6

Prediction, estimation, probability and miscellaneous mathematics content

This chapter concerns nonroutine problems that address the content areas of prediction, estimation and probability, as well as a few miscellaneous mathematics content not covered in the other chapters. Some of the problems in this chapter are indeed mathematical in content; however, the content focus represents a shift from Chapters 4 and 5 from content that has minimal practical application for most students to content that many students can apply to solving meaningful nonroutine problems in their lives, such as the nonroutine problems in the curriculum discussed in Chapters 7 to 9. For example, in this chapter, "Assigning grades" has students look at the patterns in the raw scores on a major test to assign grades to students; "Best menu" has students determine the best lunch menu for their school from the student body's perspective and "Predicting the number of lunches" has students predict the number of lunches that will be needed on a Friday in their school, all real problems that need to be solved in their school.

Introductory nonroutine problems

There are six introductory nonroutine problems that address the content of prediction, estimation, probability and miscellaneous mathematics content discussed in detail in Chapter 3, except "Dividing a line segment" which is discussed in detail in Chapter 5:

DOI: 10.4324/9781003393283-6

(1) **"Assigning Grades"**: Students assign grades given raw scores and two given grades without knowledge of the test or the total number of problems on the test.

(2) **"Number of Beans"**: Students estimate the number of beans in a large jar without touching the jar, using at least three methods.

(3) **"Most pleasing rectangle"**: Students determine which rectangle with one dimension of 6 inches will be judged most pleasing. In addition to Chapter 3, this problem is discussed in Chapter 4, Geometry.

(4) **"Marbles"**: This is an introductory nonroutine problem for which students predict the percentage of four types of marbles (different colors) in a bag. Students work in groups and are allow to take ten marbles at a time, record the information and return the marbles to the bag. A portion of the assessment is a combination of the group's accuracy (i.e., group's percentages compared to actual percentages) and efficiency (i.e., number of marbles sampled).

(5) **"Dividing a line segment"**: Students devise a method to divide a line in a manner that embodies the Golden Mean. This problem is discussed in detail in Chapter 5. Although the problem can be solved using the quadratic equation, for most students a solution involving guess-and-check is much more appropriate and likely, requiring only a minimum understanding of a variable.

(6) **"Five calculations"**: This introductory nonroutine problem involving number sense is discussed in detail in Chapter 3.

Intermediate level

There are seven problems at this level of difficulty:

(1) **"Buddha"**: Students calculate the probability of a sea turtle putting its nose through a ring randomly floating in the Pacific Ocean. The problem is based on a quote by Buddha concerning the probability of being born human.

Sample student problem statement. There's a story that once the Buddha walked with his monks by the seaside and he said to them,

> Monks, if there were a blind turtle swimming in the oceans of the world and also a wooden yoke, and this blind turtle came up for air once every hundred years, do you think, monks, that this blind turtle could put her head through that wooden yoke?

The monks said, "No, sir. That's impossible. They couldn't be in the same place at the same time if they're swimming around in the oceans of the world." The Buddha said, "No. It's not impossible. It's improbable, but not impossible." And he added, "The same improbability reigns over being reborn a human being." The task for this problem is to calculate the probability of a blind sea turtle, coming to the surface of the Pacific Ocean, putting its nose through a ring randomly floating on the surface of the Pacific Ocean. The purpose of this task is to give you practice solving a nonroutine problem involving probability. Specifically:

1   As stated above, the task is to calculate the probability of a blind sea turtle, coming to the surface of the Pacific Ocean, putting its nose through a ring randomly floating on the surface of the Pacific Ocean. You will need to make some assumptions (e.g., the size of the ring), and use some written resources to complete the problem.

2   You will submit a written report documenting how you solved the problem including false starts and your final calculation of the probability expressed as a decimal.

3   Only a few students will be asked to orally explain their project, but each student should be prepared to give an oral presentation.

Teaching suggestions. Prior to starting the task students need to know that the probability of an event is the ratio of the number of favorable outcomes over the number of total possible outcomes. Also, the students should know the need in the problem to keep measurements in the same units. It would be appropriate to review or introduce the concept of the probability of an event at the same time you introduce the problem. However, it does not seem appropriate to discuss the need to use similar units since part of the solution of the problem involves realizing the need to convert measurements to the same units. Perhaps this concept could be mentioned in a lesson within a week of introducing the problem.

Go over the directions and answer any student questions. You might want to include some background information concerning the life of Buddha. Students should be aware that they cannot calculate the probability exactly and that the task is to come up with the best possible calculation (expressed as a decimal). You might cite similar type of situations in real life (e.g., Gallup polls, calculating the size of a large crowd).

Processing. In processing this project, you certainly want to make sure that the students understood the basic concepts involved in calculating the probability of the event. Also, you might discuss how students obtained the information they needed to calculate the probability. For example, how did they determine the size of the Pacific Ocean? Were there false starts?

Suggested Format. The students work individually on this task. The oral reports, if any, and the processing will require a portion of a class period. The project can be completed in about five school days.

Note: While it is suggested that this problem be given as an individual problem, without scaffolding that addresses the need to be consistent in the use of units (for example, are the same units used for the surface area of the ocean and the size of the ring?), you can adjust the format to be at the appropriate level of difficulty for your students. For example, you might decide that for your context a short lesson on consistency of units is necessary to insure an appropriate level of difficulty.

Enrichment and Extension. A good discussion might be the difference between an improbable event and an impossible event. Also, students can generate other improbable events and calculate their probability (e.g., the probability in this galaxy [or universe!] to be on Earth). Finally, there is a formula for determining the probability of there being intelligent life elsewhere in the universe.

(2) **"Predicting the Number of Lunches":** Students predict the number of lunches that will be served on a given day at their school. They can only use data collected from students to make the prediction.

Sample student problem statement: The task for this problem is to predict the number of lunches that will be served for lunch Friday, _____. The purpose of this task is to give you practice solving a nonroutine problem involving prediction and gathering data. Specifically:

1 Each group will develop and submit a written plan explaining how you will predict the number of lunches that will be served on _____. You need to include the steps you will follow. For this problem, you are restricted to data gathered from questioning other students. For example, you cannot question cafeteria employees.

2 Your plan will be returned and you will have until the Wednesday before the prediction day to gather and analyze data to make your prediction. You will submit your prediction in writing and include how you arrived at your conclusion and why you believe it is accurate.

3 On the Thursday before the prediction day, a member of your group will be asked to make an oral presentation

(maximum: two minutes) on your prediction, how you arrived at your conclusion and why you believe it is accurate.

Teaching suggestions. Go over the directions and answer any student questions. You might want to emphasize the order of the tasks: (a) developing a written plan that you review, (b) gathering and analyzing the data, (c) completing the written report and making the final prediction, (d) oral presentations and (e) counting of the actual number of lunches served. Be sure that the students are aware that part of the grade is based on the amount of effort needed to make the prediction. In other words, ideally you want a quick and accurate method to make the prediction. The evaluation of the students' work can be based on the quality of their prediction as documented in their final written report, including two factors, accuracy and amount of effort needed to make the prediction. Was the proper data collected and well analyzed? Is adequate justification provided to support the quality of the prediction? Of course, students have to balance this goal with a concern for having enough data to feel confident in their prediction. It seems appropriate to let the students know what the lunch will be on the date for the prediction. Also, you might want to suggest to groups that they brainstorm possible difficulties they might encounter collecting data, such as making sure the questionnaires are filled out by a good sample and collected.

When you collect the written reports you might want to insure that groups have a workable plan and have determined how they will gather the data accurately. In addition, you might want to arrange to copy any questionnaires and arrange for distribution in homerooms. When the projects are complete, collect the written reports and have the groups give their oral presentations.

Processing. Besides whatever comments naturally come out of reviewing all the reports, you might discuss what type of difficulties students encountered in gathering data. Were their factors that groups did not consider fully?

Suggested Format. The students work in groups of three and are given a portion of two periods to develop their first plan, and a portion of two periods to analyze the data and prepare their final report. The oral reports and the processing will each require a portion of a class period. The project can be completed in ten to fifteen school days.

Enrichment and Extension. Each group can predict how many of each type of drink (e.g., skim milk, regular milk, orange juice, etc.) will be sold on a given day. Based on your evaluation of their work, their interest and the processing, you might want to assign a similar additional problem (e.g., How many students will ride a bus on a given day?) to give them an opportunity to further develop their skills predicting.

(3) **"Flipping Coins."** Students predict the outcome of tossing ten coins 1,000 times (e.g., how many times will the result be 6 heads and 4 tails).

Sample student problem statement. The task for this problem is to use data that will be provided to predict the outcome of flipping coins in groups of ten. The purpose of this task is to give you practice solving a nonroutine problem in probability and in interpreting data. Specifically:

1   Each group will predict the outcome of flipping coins in groups of ten 1,000 times; that is, how many times (out of 1,000) will the outcome be 10 heads, 0 tails; 9 heads, 1 tails; 8 heads, 2 tails; 7 heads, 3 tails; … .; 1 heads, 9 tails and 0 heads, 10 tails.

2   For this problem, your group will be restricted to using the data that I provide to make your prediction. Also, your group will be evaluated on not only how accurate your prediction is but also how little data you used to make your prediction. Of course, your group will have to document how you were able to make your prediction with confidence with the amount of data used.

3   Each group will prepare an oral presentation of two minutes or less on the gathered information. I will select the member of your group to present the oral report.

4   Your group will turn in a written report including your prediction, how you arrived at your prediction and why you believe your prediction is accurate.

Teaching suggestions. The task for this problem is for each group to predict the outcome of flipping coins in groups of ten 1,000 times; that is, how many times (out of 1,000) will the outcome be 10 heads, 0 tails; 9 heads, 1 tails; 8 heads, 2 tails; 7 heads, 3 tails; … .; 1 heads, 9 tails and 0 heads, 10 tails. The students are restricted to using the data sheets with which you provide them, each sheet with 100 randomly generated outcomes for 10 tosses. They are evaluated on not only how accurate their prediction is but also how little data they used to make their prediction. The students should have some basic sense concerning the likelihood of events. You may want to give the handout FLIPPING 50 COINS to be done individually before starting the problem. If the students complete the handout, it might be interesting to have each individual student prediction the outcome of the 1,000 outcomes in the nonroutine problem before looking at the data.

Give the students the problem statement and go over the directions and answer any student questions. You might want to emphasize the purpose of requiring the students to make their prediction with as little data as is mathematically sound by discussing the importance of time in practical settings. For example, a company would want to make accurate predictions using as few employee hours as possible. Ask the class if they can think of any real-life situations where you would want this combination of accuracy and speed.

Let the students know that you have 30 data sheets with 100 outcomes of flipping ten coins each and that groups can request as many sheets as they need to complete the project. You should indicate to them that you will give different groups different data sheets and you will keep record of which sheets each group was given. You might require that they request data sheets as they need them, reemphasizing the need to use as little data as possible to make an accurate prediction. You can create the data sheets using a random number generator in a simple BASIC computer program.

Students should understand that a prediction using very little data that is not justified (i.e., students do not provide any sound reasons for their conclusion) will not be given much credit, even if accurate. When the projects are complete, have the groups give their oral presentations.

Processing. Besides whatever comments naturally come out of reviewing all the reports, discuss how students knew when they had enough data. For example, one way of doing the problem is to record data until it becomes clear that the data is not changing significantly. For this project, that might mean that the two sides of the data are very balanced. For example, the number of outcomes of 7 heads, 3 tails is very close to the number of outcomes of 3 heads, 7 tails, or the percentage of each type of outcome changes vary little with additional data. The significance of these approaches is that if the student tries something (recording data), a pattern (requiring no knowledge of difficult mathematical content) will eventually emerge and make the problem much easier to complete.

Suggested Format. The students work in groups of three and are given a portion of two periods to work on the project and prepare their oral report. The oral reports and the processing will each require a portion of a class period. The project can be completed in five to ten school days.

Enrichment and Extension. A good extension is to predict the outcomes of flipping groups of 12 coins. It would be interesting to have students predict the outcome before generating data. In addition, if appropriate for your class, you can demonstrate the formula and discuss how similar the results are to their calculations.

Flipping 50 coins

Experiment #1

Equipment Needed: 1 coin; Prediction Sheet #1 (below).

Directions: This is a simple experiment. You will be asked to flip a coin 50 times and record the results. Here are the directions:

(1) Before you start flipping the coin, fill out Prediction Sheet #1.
(2) Flip a coin 50 times (be sure that the coin turns at least 3 or 4 times in the air) and record the results of each flip.

Prediction sheet #1

Remember, you will be flipping a coin 50 times!

(1) How many times do you think you will flip a head (out of 50)?_____
(2) How many times do you think you will flip:
    (a) exactly two heads in a row?_____
    (b) exactly three heads in a row?_____
    (c) exactly four heads in a row?_____
    (d) exactly five heads in a row?_____
    (e) exactly six heads in a row?_____
(3) Check (x) each of the following which you think are likely to happen when you flip the coin 50 times:
    _____(a) You will flip more heads than tails.
    _____(b) At least once, you will flip exactly three tails in a row.
    _____(c) You will flip between 23 and 27 heads.
    _____(d) You will flip between 20 and 30 tails.
    _____(e) You will flip exactly 25 heads.
    _____(f) You will flip between 15 and 35 tails.
    _____(g) At least twice, you will flip exactly two heads in a row.
    _____(h) At least once, you will flip exactly six tails in a row.

(4) **"The Best Menu for the School."** Students predict the most popular menu for the school based on a questionnaire they develop and administer to a portion of the student body.

Sample student problem statement. The task for this problem is to identify the lunch menu for five days that would be most liked in this school. The purpose of this task is to give you practice solving problems using a questionnaire as a tool and requiring persistence. Specifically:

1   Each group will prepare a list of five lunches that would be most liked in this school.

2   In the first part, your group will brainstorm ideas for the list and, based on those ideas, prepare a list of tentative lunches and a questionnaire to gather information from students to determine whether your ideas are good. I will copy and distribute the questionnaires and return the completed questionnaires to your group.

3   In the second part, you will analyze the data from the questionnaires and prepare your final list. To prepare your final list, your group may construct and administer a second questionnaire as long as the questionnaire addresses items on your first list or new items not on other group's original list. Your group's final list can only contain ideas on your original list or new ideas not on any other group's original list.

4   Each group will prepare an oral presentation of two minutes or less on your work.

5   Each group will hand in a written report including the group's final list and documentation of the group's process of selecting the final list.

Teaching suggestions. This problem can be introduced by telling students that this problem is similar to what a supermarket or restaurant would need to determine how to avoid ending up with food that could not be sold or not having enough food. Go over the directions and try to answer any questions. Be sure that students understand the two stages of completing the project.

Be sure students realize that they can include more than five lunches on their questionnaire. One option is to give the students a format for their lunches such as one entree that includes at least three foods (e.g., hamburger, salad, and peas) and one dessert. I would recommend not including a drink. When the projects are complete have the groups give their oral presentations. When the presentations are complete, collect the written reports.

Processing. A major focus of the processing could be the role of the two questionnaires in solving the problem. Discussion of examples of how the data from the questionnaires influenced the final lists could be fruitful, particularly examples in which the data was counter to the intuition of the groups.

Suggested Format. The students work in groups of three and are given a portion of two periods for the first part and an additional portion of one period for the second part. The processing and oral presentations should each require a portion of a class period. The project can be completed in 10 to 15 school days.

Enrichment and Extension. A good example of enrichment would be to have a speaker from a food market or restaurant to speak about how their business determines how much food to order.

(5) **"Leaves on a Tree."** Students calculate the number of leaves on a large tree. They are required to develop at least three methods.

Sample student problem statement. The task for this problem is to estimate the number of leaves on a tree specified by the teacher. Your group can use any method as long as the method does not cause damage to the tree. This task gives you practice developing and evaluating a number of alternative methods to solve a problem. Specifically:

1   Your group will develop and evaluate at least three methods to estimate the number of leaves on the tree. You are required to determine your final estimate.

2   Your group will prepare an oral presentation of two minutes or less on the gathered information. I will select the member of your group to present the oral report.

3   Your group will turn in a written description of: the process you used to arrive at your final estimation, why you believe your estimate is valid and a description of all other methods you considered.

Teaching suggestions. For this problem you need to establish which tree, preferably on the school property, is to be used for the project. Obviously, a large tree lends itself well to this project. Go over the directions and try to answer any questions. It might help to emphasize that a major portion of the problem is to come up with a variety of methods; that is, a group that finds one method that is very accurate will not do as well as a group that finds a few good to very good methods. If feasible, it would be helpful to allow some class time outside actually inspecting the tree. It is certainly reasonable to expect students to use some time outside class at the tree. For this problem, there is not a practical method to verify the accuracy of the estimates. It should be clear to students that the accuracy of their methods will be judged by the opinion of the students and a panel of teachers (or just you).

When I first considered this problem before inspecting a tree, I wondered how could one make a good estimate! However, when I actually inspected a tree, I began to notice patterns in the distribution of the leaves which allowed me to count (or estimate) the number of leaves in a small section of the tree that then permitted me to make a reasonable estimate of larger sections of the tree, and so on. An experience that facilitated me appreciating the step of trying something!

Processing. A good focus for the processing could be for the students to get an appreciation of the wide variety of good methods to estimate the number of leaves. Therefore, it would be appropriate to select one or two good methods from each group to discuss. Another focus might be what assumptions caused some of the methods to result in poor estimations. The requirement of using at least three methods facilitates the students' understanding the step of persistence; however, in practice it is common that students with little experience with nonroutine problems use three methods as required but they do not address big discrepancies in the three estimates, which is inconsistent with the essence of the step of persistence. I remember a group for which there was a difference of over 200,000 in the estimates and the students' estimate was just the average of the estimates! Therefore, a good focus of the processing is how students implemented the requirement of three methods and whether their implementation was consistent with the essence of the step of persistence. For example, did they address discrepancies in the estimates?

Suggested format. The students work in groups of three and are given a portion of two periods to discuss their strategy and "inspect" the tree (if feasible). The processing and oral presentations should each require a portion of a class period. The project can be completed in four to eight school days. This problem can also be assigned as an individual assignment.

Enrichment and Extension. One extension is to estimate the number of leaves on trees in your community, state or the entire United States. Also, a similar problem such as estimating the number of blades of grass on a lawn might be appropriate.

(6) **"Favorite Television Shows."** Students predict the favorite television shows of students through the use of a questionnaire.

Sample student problem statement. The task for this problem is to identify the five television shows most liked in this school. The purpose of this task is to give you practice solving problems using a questionnaire as a tool and requiring persistence. Specifically:

1   Each group will prepare a list of the five television shows most liked in this school. Your work on this task will be in two parts. In the first part, your group will brainstorm ideas for the list and, based on those ideas, prepare a list of tentative shows and a questionnaire to gather information from students to determine whether your ideas are good. I will copy and distribute the questionnaires and return the completed questionnaires to your group.

2   In the second part, you will analyze the data from the questionnaires and prepare your final list. To prepare your final list, your group may construct and administer a second questionnaire as long as the questionnaire addresses items on your first list or new items not on other group's original list. Your group's final list can only contain ideas on your original list or new ideas not on any other group's original list.

3   Each group will prepare an oral presentation of two minutes or less on your work.

4   Each group will hand in a written report including the group's final list and documentation of the group's process of selecting the final list.

Teaching suggestions. This problem can be introduced by telling students that this problem is similar to a question television networks need to answer, What are the favorite television shows for our audience? Go over the directions and try to answer any questions. Be sure that students understand the two stages of completing the project. It might help to indicate to students the need for a questionnaire; that is, the questionnaire will provide data to judge whether the ideas that the group believes are good are, in fact, good in the opinion of other students. Citing examples, such as screenings for potential ads and polls on certain issues, might clarify the need for this step.

Typically, this nonroutine problem is later in the curriculum; therefore, students should appreciate the purpose of the second questionnaire, to clarify unclear results of the first questionnaire and to gather information about ideas raised by the data from the first questionnaire. The second questionnaire is connected with the need for persistence in problem solving. You may want to either alert them to this issue during the project or wait until the processing to discuss whether groups chose to give a second questionnaire and how that decision affected the quality of their results. When the projects are complete, have the groups give their oral presentations. When the presentations are complete, collect the written reports.

Processing. A major focus of the processing could be the role of the two questionnaires in solving the problem. Discussion of examples of how the data from the questionnaires influenced the final lists would be fruitful, particularly examples in which the data was counter to the intuition of the groups and/or examples that illustrate good or poor understanding of the step of persistence.

Suggested Format. The students work in groups of three and are given a portion of two periods for the first part and an additional portion of one period for the second part. The processing and oral presentations should each require a portion of a class period. The project can be completed in ten to fifteen school days.

Enrichment and Extension. A number of similar project are possible such as favorite radio stations, favorite albums and differences between parents' and students' taste in television.

(7) **"Correlation."** The problem is to develop a method for assigning a number to indicate the degree of similarity between two rank orders. The method should discriminate between rank orders that are similar, are dissimilar and have no relationship, as well as being able to discriminate between rank orders in the same category.

Teaching suggestions. Give the students the attached handout including the sample data (see below). Using the sample data, be sure that the students intuitively understand the difference between similar rank orders, dissimilar rank orders and rank orders that have no significant relationship. One option is a short quiz after the explanation to classify three examples. If the students have demonstrated an ability to generate a variety of solutions and refine their solutions in solving other nonroutine problems then the handout and above directions should be adequate. However, if this is not the case, students are likely to be satisfied with a solution such as summing the differences of the rank orders for each of the ten ranks and letting a low sum indicate similarity and a high sum dissimilarity. If you expect that type of response from your students, an option is to present the class with the above solution as an example of one solution and challenge them to come up with a better solution.

Solutions. Some students take a sum of the differences and divide by a number to make the scale convenient. Others square the differences to establish a wider spread for differences. The rank correlation formula,

$$r = 1 - \frac{6\sum d^2}{n(n^2 - 1)}$$

where d = the difference between two ratings and n = the number of items ranked can be introduced for discussion. In processing, I suggest to students that it is unlikely that you would recall this formula; however, the methods you devised probably give very similar results and did not require you to memorize any formula(s). For example, the nonroutine problem "Grading" demonstrates that some problems that can be solved with somewhat complicated formulas can be solved well using more simple methods that do not depend on you remembering and implementing a complicated formula.

Processing. If time, you may want to compare a few (or even one) of the student's methods with the results of the actual formula, pointing out that similar results indicate the power of their solutions which they developed and are probably easier to use than the formula.

	Similar			Dissimilar				No Relationship		
1	2	2	1	10	1	9	10	5	7	6
2	3	1	2	9	10	8	9	6	5	2
3	4	4	3	8	9	7	8	2	1	8
4	5	3	6	7	8	6	7	6	9	4
5	6	6	5	6	7	5	4	3	3	10
6	7	5	4	5	6	4	5	10	6	1
7	8	8	7	4	5	3	6	7	2	5
8	9	7	8	3	4	2	3	1	10	9
9	10	10	9	2	3	1	2	4	3	5
10	1	9	10	1	2	10	1	9	4	7

Advanced

There are two problems at this level of difficulty:

(1) **"Approximating π."** Students calculate π in ten different ways. Students are evaluated on accuracy, creativity and variety. This nonroutine problem is fully discussed in Chapter 4, Geometry; however, this is an excellent problem later in the mathematics curriculum when students have been exposed to a variety of formulas involving π.

(2) **"Making Predictions":** Students predict the outcome of certain experiments involving picking objects out of a bag containing

20 green marbles, 10 red marbles and 5 blue marbles (e.g., probability of picking two green objects in a row).

Sample student problem statement. The task for this problem is to predict the outcome of certain events involving a bag containing 20 green marbles, 10 red marbles and 5 blue marbles. The purpose of this task is to give you practice solving a nonroutine problem involving probability and prediction. Specifically:

1 Your group will predict the probability of the following five events: given a bag with 35 marbles of which 20 are green, 10 are red, and 5 are blue, what is the probability of picking two consecutive green marbles; picking two consecutive red marbles; picking two consecutive blue marbles; picking three consecutive green marbles and picking ten consecutive green marbles. Express all your answers as decimals (rounded off to the nearest thousandth, if necessary).

2 Each group will prepare an oral presentation of two minutes or less on the final method you used to calculate your answers. I will select the member of your group to present the oral report.

3 Your group will turn in a written report including a description of your final method, why you selected that method and a description of at least two other methods you considered.

Teaching suggestions. This problem would not be appropriate for students that know the formula for calculating the probability of these events. The problem could be adjusted to require them to calculate the formula from experimentation rather than by using a formula. Go over the directions and answer any student questions. One recommended option is to make available to each group a bag of 35 marbles as described. Of course, this makes the problem easier in that you are helping them decide what to try; however, without that scaffolding the problem may be too difficult. As an advanced problem, this is meant for groups that have already had good experience solving nonroutine problems. This is a good project to require a progress report. When the projects are complete, have the groups give their oral presentations. When the presentations are complete, collect the written reports.

Processing. Besides whatever comments naturally come out of reviewing all the reports, it seems important in the processing to discuss how students gathered data. It would be valuable to compare two similar approaches in which one group found a better variation due to persistence.

Suggested Format. The students work in groups of three and are given a portion of two periods to work on the project and prepare their oral report. The oral reports and the processing will each require a portion of a class period. The project can be completed in five to ten school days.

Enrichment and Extension. A discussion of the relevant formulas would probably be of interest to students after completing the project.

7

Introductory nonroutine problems in a non-mathematical context

Chapters 7 to 9 concern additional nonroutine problems that require the three steps of a nonroutine problem to solve; however, the content focus of the problems is not typically considered to be mathematical. A major objective of the curriculum is that students will be able to apply the three steps in solving a mathematical nonroutine problem to solve meaningful problems that might not appear to be mathematical in nature, but require the three steps for a good solution. As previously mentioned, field-testing indicates that to insure transfer to the student's day-to-day life most students need specific experience solving meaningful nonroutine problems in areas not typically considered "mathematical." In addition to facilitating transfer of the ability to solve nonroutine problems in mathematics to other meaningful contexts, these problems address important issues and problems that lend themselves to the three steps of solving a nonroutine problem and engage students in meaningful problem solving. Specifically, the chapters cover the following four strands, introductory nonroutine problems in a nonmathematical context (Chapter 7), developing ecological awareness and sustainability (Chapter 8), increasing appreciation of diversity (Chapter 8) and individual and class generated nonroutine problems (Chapter 9). These four strands involve approximately 40% of the problems suggested in the curriculum.

DOI: 10.4324/9781003393283-7

Introductory problems

There are five problems, all at the introductory level, in this chapter. The first two, "Scavenger hunt" and "Researching a topic," are not necessarily non-routine problems, but rather problems that give the students experience with researching topics, a necessary skill for some of the nonroutine problems in Chapters 7 to 9. The remaining three nonroutine problems have the additional characteristic of tending to create a positive learning community in the classroom.

(1) **"Scavenger Hunt":** Students use a variety of resources to gather certain facts or bits of information (e.g., What size is the largest centipede ever found?). Students must use written, electronic and people resources.

   Sample student problem statement: The task for this problem is to use a variety of resources to answer as many of the questions on SCAVENGER HUNT: QUESTIONS TO ANSWER as possible. This task gives you experience using a variety of resources to gather information. Specifically:

   1   Your group will hand in a written report that will include your answers to the questions and how you found the answers, including methods you tried that were not successful. For each answer, be sure to include the source for your answer.

   2   Each group will prepare an oral presentation of two minutes or less on your answers to questions 11 to 15. I will select the member of your group to give the oral presentation.

   Teaching suggestions: This problem focuses on the skill of using resources to find information, a skill essential to solving many of the nonroutine problems. Each group of students answers 15 questions that require them to use a variety of resources. Ten of the questions require them to discover the information that correctly answers the question and five of the problems require them to find interesting information on a specified topic. This problem is truly introductory in that it provides students with experience researching questions using a variety of resources, including interactions with other people, a skill that they will need for other nonroutine problems in Chapters 7 to 9. Students need to know how to use basic tools to research questions, including interactions with adults. Other nonroutine problems in this curriculum assume this basic skill. In addition, students should know how to access other general resources the library offers (e.g., computer services). An orientation by the school librarian might be appropriate. Let the students know that the primary purpose of this problem is to give the students practice with the basic skills involved in gathering data and information from both written and people resources and that this is a skill that they will need to use in many of the other nonroutine problems in the curriculum.

Go over the directions and try to answer any questions. It might help to emphasize that you expect the students not only to answer the questions, but also to document their attempts finding the information, including unsuccessful attempts. Students should understand that the evaluation of their attempts will be based on their ability to document them in a clear, concise way – not on whether they had false starts or not. Perhaps it would be useful to add that most of the nonroutine problems will naturally involve false starts and that in the processing of many problems a portion of the discussion will focus on how the false starts are a natural part of the process and how they can be used to generate good solutions. If you can give the students an interesting example of information you gathered, it would probably give the students a better sense of what is required for questions 11 to 15.

Comment. The questions asked on SCAVENGER HUNT: QUESTIONS TO ANSWER are generic questions applicable to most, if not all, high schools. You may want to substitute or add questions more relevant to your community. The questions below were used sometime between 1992 and 1995, before the internet was used extensively for this type of information gathering; therefore, you might want to update the questions from one to ten.

When the projects are complete, have the groups give their oral presentations. When the presentations are complete, collect the written reports.

Evaluation and Processing. Two important concepts to focus on in the processing are successful and unsuccessful methods of using resources to gather information and the role of false starts in problem solving. One method to focus on successful and unsuccessful methods of using resources is to find one example of each from each group report and discuss the examples as a class, perhaps generating a list of what worked and what did not work. To focus on the role of false starts it would help to find a few examples from the written reports in which the significance of false starts was documented (e.g., if a group wrote "Although this did not give us the answer, it gave us the idea to

…. which resulted in the correct answer"). Discussing these examples would also reinforce the importance of documenting false starts.

Suggested Format. The students work in groups of three and are given a portion of two periods to discuss their strategy. It seems appropriate for this problem that students gather their information outside class-time (e.g., go to the library). The processing and oral presentations should each require a portion of a class period. The project can be completed in four to eight school days.

Enrichment and Extension. Each group can generate three additional questions to answer. Groups can be evaluated on their effectiveness in answering other group's questions and the ineffectiveness of other groups answering their questions. Of course, questions would need to be screened by the teacher for appropriateness, particularly whether the question is reasonably possible to answer.

Below is a sample of possible questions; however, as mentioned, questions more appropriate for your community or the types of data now accessible should be substituted.

Scavenger hunt: questions to answer

1. Which teacher at your high school has been teaching the most years?
2. How much did a first class stamp for a regular size letter cost in 1959?
3. How large was the first graduating class at your high school?
4. Who was President of the United States in 1895?
5. What is the size of the largest centipede in the world?
6. Who was valedictorian of your high school in 2000?
7. Identify two students in your high school that have the same birthday.
8. When was the bicycle invented?
9. What is the total school budget for your school district this year?
10. Who was Guiseppe Garabaldi?
11. Find an interesting bit of information concerning the 1986 World's Fair.
12. Find an interesting bit of information concerning the 2000 graduating class of your high school.
13. Find an interesting bit of information concerning the 1997 professional baseball season.

14 Find an interesting bit of information concerning a retired teacher from this high school.

15 Find an interesting bit of information concerning Finland.

Answers to the more difficult questions: (1) the cost of a stamp in 1959 was four cents (*The Statistical Abstract of the United States*, 1991, p. 555); (2) the largest centipede is 13″ long and 1.5″ breadth (*The Guinness Book of Records*, 1992, p. 103); (3) the first bicycle was probably invented in France about 1790 and the first bicycle with a steerable front wheel was invented in either 1816 or 1817 (the answer to this question is not as absolute as the other questions; source: two encyclopedias); and (4) Guiseppe Garibaldi lived in Italy from 1807 to 1882 and is primarily known for his attempts to unite Italy (encyclopedia).

(2) **"Researching an interesting topic":** Using written references including articles, books, internet and interlibrary loans, students research a topic in which they are interested.

Sample student problem statement: The task for this problem is to find information from written sources about a topic in which you are interested. This task will teach you how to find information about a subject in which you are interested. Specifically:

1 You will pick a topic you are interested in researching. You will have to find information from at least one article from a magazine available at your local library, one article from a magazine obtained through interlibrary loan, and one book.

2 You will hand in a written report in which you identify at least five new ideas or bits of information you learned about your topic. You will list all the sources you used for your research, including in which resource you found each bit of information.

3 Only a few students will be asked to orally explain their project, but each student should be prepared to give an oral presentation.

Teaching suggestions: This problem focuses on written resources, a resource I believe students should be familiar with in addition to resources from the internet and people interactions. Of course, if you believe this is not an important resource for your students at this time, then this problem can certainly be omitted or adjusted (like the previous problem) to fit your present context or goals. Similarly, you can consider whether the inclusion of obtaining a resource from interlibrary loan is appropriate for your students and context.

For this problem, you need to decide how much support the students need to be able to successfully complete the project. For example, if you believe students' research skills are minimal, you might want to schedule an orientation to finding resources in your school's library. You might schedule one class in the library in which students can start their research with the aid of the librarian and you.

Go over the directions and answer any student questions. You might want to have students submit their topics to you for approval before they start their research. Also, let students know that if they have difficulty with their topic, they can ask you for permission to change their topics. Since this might be the first individual project, you might require a progress report about half way through the project. Since it is an individual project, I would suggest that you allow a period of six to ten school days to work on the project.

Evaluation and Processing. When all the projects are complete, you might take time to share selections from each student's report, including any interesting resources they used. Perhaps you can ask some students to give a brief oral report. You may ask students whether they felt the project was valuable to them. Do they think they will use the skills they learned on their own in the future?

Enrichment and Extension. As mentioned previously, this problem and "Scavenger hunt" focus on basic researching skills the students need for later nonroutine problems; therefore, you need to insure that after the two problems you feel confident that they have those skills. If not, additional problems would be appropriate. Students can research additional topics of interest. Or students can generate a list of a few topics that many of the students are interested in learning about (e.g., a popular musical group) and research those topics in groups.

(3) **"Awesome Playlist":** Students develop a half hour playlist of music that the class will like.

The task for this problem is to create a 30-minute playlist that the students in this class will like. The purpose of this task is to give you practice solving a nonroutine problem involving the use of a questionnaire. Also, the task will help the class get to know each other better. Specifically:

1   Each group will create a 30-minute playlist of music which you believe the class will like. It is suggested that each group start by brainstorming ideas for the playlist.

2   Your group will be able to develop and give out a questionnaire that can be answered in ten minutes or less to gather information. Any song not mentioned on your questionnaire and mentioned on another group's questionnaire cannot be used on your playlist.

3   Each group will prepare an oral presentation of five minutes or less to convince the class that your playlist is the best. You should prepare a list of the songs on your playlist for students. You can play portions of your playlist as long as it is within the five-minute presentation. I will select the member of your group to present the oral report.

4   In addition, your group will turn in a written report including a list of the songs on your playlist with times and a description of the process your group used to arrive at your final playlist.

Teaching suggestions: Give the students the problem statement and go over the directions and answer any student questions. Indicate any restrictions on what students can include on the playlist (e.g., inappropriate sexual content). You might want to make sure all groups have access to needed equipment. You might want to have students give out their questionnaires on the same day. The questionnaires can be given as a homework assignment to save class time.

The oral presentations are restricted to five minutes to save class time and, also, to force students to decide what is the most relevant information to present. You can point out to students the similarity in the process of preparing a short commercial for a product. On the other hand, if you can allow more time for presentations, it may help class atmosphere. You should indicate to students what the result of the project will be. For example, will students be able to play the playlist during certain group activities? Will the best playlist be played once (and once only) during class? Will the playlist be available for copying?

When the projects are complete, have the groups give their oral presentations. When the presentations are complete, collect the written reports.

Evaluation and Processing. Besides whatever comments naturally come out of reviewing all the reports, discuss any surprises that groups encountered in gathering data, particularly if the surprise supports the need to gather data to solve a problem.

Suggested Format. The students work in groups of three and are given a portion of two periods to develop their questions, and a portion of two periods to analyze the data and prepare their oral report. The oral reports and the processing will each require a portion of a class period. The project can be completed in six to ten school days.

Enrichment and Extension. A good extension is to develop as a class a playlist using the data from all the groups. The process of making sense out of the combined data of all the groups would be an excellent opportunity to work on a more complicated problem. Certainly, the data, even when summarized, would require an additional questionnaire to effectively evaluate.

(4) **"Collage":** Students create a collage that will interest the students in the class. The problem also helps to build a good atmosphere in the class.

Sample student problem statement: The task for this problem is to develop a collage for this class that other students will like. The purpose of this task is to solve a nonroutine problem that requires you to gather information from others. Specifically:

1 Each group will create a collage for this class that the other students will like. The collages will be rated by the students in this class, a high rating indicating that the students would very much want the collage displayed in this class.

2 Each group will start by preparing a questionnaire requiring no more than ten minutes to complete. The questionnaire should consist of questions that you believe will indicate to your group what type of items the class would like on the collage. The questionnaire needs to be typed or written neatly in black ink.
I will copy the questionnaire and give it to the class. You will be permitted to collect additional data; for example, if you wish to develop a second questionnaire, I will give the questionnaire to the class.

3 Your group will complete the collage and prepare an oral presentation of two minutes or less explaining how you created the collage. I will select the member of your group to present the oral report.

4 Your group will turn in a written report explaining the steps your group took preparing the collage.

Teaching suggestions: The students should have some experience developing questionnaires (e.g., from other nonroutine problems). Go over the directions and answer any student questions. You should specify a maximum size for the completed collage. You might want to emphasize the purpose of the questionnaire, to gather information to help determine what other students might like on the collage. You can compare the task to the problem of creating an advertisement that would appeal to a certain group. If there are any restrictions on what material students can include on the collage, you should try to make the restrictions clear at the beginning of the project. You might suggest to students to check with you if they are not sure of the appropriateness of some material. Indicate to students what will happen to the completed collages. For example, you might display all of the collages for a week and keep one or two displayed for a longer portion of the school year. In my experience, displaying the collages helped create a good learning environment

When the projects are complete, have the groups give their oral presentations. When the presentations are complete, collect the written reports.

Evaluation and Processing. Besides whatever comments naturally come out of reviewing all the reports, explore the role of the questionnaires in the problem. Did they provide useful information? If not, why? Did a second questionnaire help any group? Also, it could be valuable comparing the project to how a company might go about preparing an advertisement.

Suggested Format. The students work in groups of three and are given a portion of three periods to work on the project. The oral reports and the processing will each require a portion of a class period. The project can be completed in six to ten school days. You should be sure that any group which wants to develop a second questionnaire has the opportunity to do so.

Enrichment and Extension It could be interesting to have a person who develops advertisements professionally to talk to the class and, perhaps, comment on the developed collages and how they might have been developed professionally.

(5) **"Getting Acquainted":** After completing a questionnaire gathering interesting information about them, students develop a similar questionnaire to gather interesting information about their teacher.

Sample student problem statement: Your group task for this problem is to gather interesting information about me, your teacher, by means of a questionnaire. The purpose of this task is to introduce you to the process of gathering information and to help us become better acquainted. Specifically:

1   Each group will develop a questionnaire for me consisting of a maximum of seven questions, each of which asks a question that you think will result in interesting information about me.

2   After I have selected three of the best questions from each group and responded in writing to those questions, each group will prepare an oral presentation of two minutes or less on the gathered information. I will select the member of your group to present the oral report.

3   Your group will turn in a description of the process you used to select your questions. The description should not exceed two

pages and should include information such as consultations with people outside your group, the method your group used to generate questions and your rationale for retaining or eliminating questions from your completed questionnaire.

Teaching suggestions: Let the students know that this problem will introduce them to the process of gathering and analyzing data through the use of a questionnaire and help the students in the class get to know each other better. Inform students that first they will answer a questionnaire given by you and then develop a similar type of questionnaire to give to you, the teacher. Indicate that, because their questionnaire will be responded to by only one person (the teacher), the questionnaire will be relatively easy to analyze compared to questionnaires for problems they will encounter later in the year. Give the students the worksheet GETTING ACQUAINTED: STUDENT INFORMATION as a homework assignment. Explain that you will select a little information from each student's assignment to share with the entire class. Let the students know that your purpose is to demonstrate how a questionnaire can be used to gather interesting information and information that also might help students (and the teacher) become better acquainted. Finally, indicate to the students that they will have the opportunity in their groups to develop a similar questionnaire to be given to you. You might want to provide an option for students who feel uncomfortable with answering the questions in the assignment.

After the assignments have been collected, select one or two bits of information about each student to share with the class. Try to select information that would interest students and be information that they would not have been likely to discover without this assignment. Discuss the results with the students by asking questions such as: Were you interested in what you learned about your classmates? Can you share one bit of information about a student that particularly interested you? Which questions were good? Can you generate better questions? You might also discuss the limitations of this type of method to get to know people (e.g., asking everyone the same questions; not a natural way of getting to know a person).

Give the students the problem statement for GETTING ACQUAINTED, go over the directions and answer any student questions. You might want to emphasize the following: (a) The problem is to determine which questions are likely to result in information about the teacher that will be of interest to the class as a whole (versus just your group). (b) The evaluation of each group will be based on your determination of effectiveness of the group in generating questions likely to result in interesting information. (c) Each group can check with you concerning the appropriateness of any question. If this is their first problem, it may be useful to collect each group's questions and descriptions, write comments and general suggestions on their papers and allow them to revise their work before you respond to the questions.

When the projects are complete, have the groups give their oral presentations. When the presentations are complete, collect the written reports.

Evaluation and Processing. It seems essential in the processing of this project to help the students see the connection between the project and problem solving. Perhaps the most effective way to do that is to focus the discussion on your evaluation of their questions and the connection with problem solving skills. For example, for each group, you might select one question that you rated high and one you rated low and discuss your ratings in terms of evidence of (or lack of) problem solving skills as a group. Also, if some groups used people outside the groups to help generate questions, you might discuss the effectiveness of that method or ask why other groups did not use outside resources.

Comment: I have found this nonroutine problem to be an excellent way to start the process of establishing a strong classroom environment for learning. I get to know the students better and they get to know me much better. There is enough structure in the lesson to prevent inappropriate information being shared.

Suggested Format. The students work in groups of three and are given a portion of two periods to develop their questions and a portion of two periods to analyze the data and prepare their oral report. The oral reports and the processing will each require a portion of a class period. The project can be completed in six to ten school days.

Enrichment and Extension: Each group can develop and administer a questionnaire for the class similar to GETTING ACQUAINTED: STUDENT INFORMATION. You may want to limit each group's questions to three.

Getting acquainted: student information

Directions: Answer at least three of the following questions on a separate sheet. Do not write anything that you would not want read to the class. Explain each answer. Each answer should be at least a couple paragraphs.

1   If you could invite three people, living or dead, to dinner and an evening of conversation, whom would you invite?
2   If pay was not a factor, what job would you most want when you finished your education?
3   If you had one year off after high school and money was not a factor, what would you do during that year?
4   What is the best gift you have ever been given?
5   What is the one thing you would like to do to make the world a better place?
6   If you could develop exceptional talent in one area, what would that area be?
7   What is the one accomplishment in your life of which you are most proud?

8

Ecology problems and problems that attempt to increase the student's appreciation of diversity

Introduction

Two major strands in the four-year curriculum focus on eight (two a year, for four years) problems with an emphasis on ecology, and four problems (one a year) that focus on increasing the student's appreciation of diversity. Hopefully, these problems directly affect the student's lives and/or are meaningful. For example, (1) Home and the Environment: Students determine how to make their homes more ecologically sound, (2) "Finding a speaker": Students find an inexpensive, interesting speaker on the topic of improving appreciation of cultural diversity, and (3) "School and the Environment": Students determine ways to make their school more environmentally sound. Each of the 12 problems requires students to explore a number of resources including written and verbal (e.g., talk to appropriate knowledgeable people) resources; evaluate the quality of the gathered data and determine which ideas are most valuable; give a clear rationale for choices; and take some action. The last problem on ecology concerning the concept of profit is more advanced, as well as more abstract, probably appropriate for advanced mathematics students with some experience solving nonroutine problems.

A common observation in the written reports for these problems concerned the perceived value of applying the three steps of a nonroutine problem to solve problems in these strands. For example, one student commented,

DOI: 10.4324/9781003393283-8

> When I first thought about this assignment [Home and the Environment], I had a bad attitude towards doing it. I thought my house was already environmentally sound, but I soon found out that there is always more that can be done.

A group of students that worked together on the project reported, "We did not realize how many things we can do to create a more ecologically sound home."

The quality of ideas students generated consistently surprised me. Due to space limitations I will only mention two. One group identified a group that was shredding out of circulation banknotes and mixing them with sludge from a nearby potato chip factory to create rich, loamy compost. In addition to uncovering a number of clever approaches that others had developed, students also came up with their own ideas based on their research and discussions with appropriate people. For example, one group suggested to arrange nature hikes and bike tours through our town, a beach community, during the summer to reduce the amount of pollution from traffic.

In my opinion, the problems in the two strands serve two important roles in the curriculum. First, they play an important role in helping the students not only understand the process of solving mathematical nonroutine problems, but also how to transfer the process of solving mathematical nonroutine problems to solving significant problems in their own lives. Early field-testing of the curriculum with just mathematical problems clearly indicated the need for exposing most students to problems in a variety of contexts to facilitate transfer. Second, in my opinion, the strands helped students understand more deeply principles of environmental education and appreciating diversity through the exploration of problems specific to their life and community. The nonroutine problems in this chapter could be labeled as generic in the sense that they could probably be effectively implemented in most communities, with results varying in each community depending on the context and characteristics of the community; therefore, the problems are appropriate for this book. However, I would add that nonroutine problems that specifically relate to your community and students would be even more appropriate and meaningful. For example, some mathematics classes have studied the distribution of toxic waste dumps and found a disproportionate number in poor communities and the students took actions to attempt to address the issue. One more example: For my classes in a primarily white middle-class community, when we planned trips to New York city that were enjoyable, educational and inexpensive, many times an unplanned outcome was an increased awareness and appreciation of diversity naturally "embedded" in the diversity of the neighborhoods we visited and our processing of the trip

Seven of the eight problems in the ecology strand are:

(1) **"Home and the Environment":** Students determine how to make their homes more ecologically sound.

Sample student problem statement: The task for this problem is to determine ways to make your home more ecologically sound. The purpose of this task is to give you practice solving a nonroutine problem affecting the environment and your life. Specifically:

1 For this problem, you will research at least five (and a maximum of ten) ways to make your home more ecologically sound. You will probably need to use a variety of resources, including written and people resources.

2 You will turn in a written report including a final list of ways to make your homes more ecologically sound (minimum of five ways, maximum of ten) and a description of how you arrived at your final list. In addition, if your family has already adopted ways to make your home more ecologically sound, you may include those in your written report, but not in your final list, these former actions will affect your evaluation in that students whose family has already adopted a variety of ways to make your homes more ecologically sound will have these past actions considered in the evaluation of their work.

3 Only a few students will be asked to orally explain their project but each student should be prepared to give an oral presentation.

Teaching suggestions: Prior to starting the task, students need to have experience using a variety of resources to gather information. Go over the directions and answer any student questions. This problem requires the students to use a variety of resources to complete the project successfully. You need to judge how much help it is appropriate to give the students. For example, many good ideas or leads can come from talking to appropriate people (or organizations) in the community. Do you think that at least some of your students would use people resources without your guidance? If not, you may want to suggest to students that they think of at least a few people they might talk to for ideas. Also, you need to decide if there is a need for you to guide them in terms of written resources available. This project is an appropriate one to require a written progress report about halfway through the project.

Processing. A good focus for the processing is what resources people used to complete the project. Selecting and discussing a variety of resources students used is a good way to focus the discussion.

If there are resources that students failed to use, it would be useful to discuss why these were not used. Also, drawing student attention to how this problem can directly affect their lives might increase their appreciation of the problem solving techniques. As part of the project, you might ask students if they are willing to commit themselves to implementing at least a few of the methods; if yes, you can tell the students that in two weeks you will ask them for an update on the changes.

Suggested Format. The students work on this project individually; therefore, only a portion of one class period is required to introduce the problem and one class period to process the problem. The project can be completed in five to ten school days.

Enrichment and Extension. The class can compile the suggestions, perhaps research additional methods, and make a handout, booklet or internet resource that would be made available to interested people. Collaborating with a community organization could be a valuable experience.

(2) **"School and the Environment"**: Students determine ways to make their school more environmentally sound.

Sample student problem statement: The task for this problem is to determine ways to make your school more ecologically sound. The purpose of this task is to give you practice solving a nonroutine problem affecting the environment and your life. Specifically:

1   For this problem, each group will research at least five (and a maximum of ten) ways to make this school more ecologically sound. You will probably need to use a variety of resources, including written and people resources.

2   Each group will prepare an oral presentation of two minutes or less on the gathered information. I will select the member of your group to present the oral report.

3   Your group will turn in a written report including a final list of ways to make this school more ecologically sound (minimum of five ways, maximum of ten) and a description of how you arrived at your final list and what resources you used, including resources that didn't result in any new ideas.

Teaching suggestions: As in the previous problem, prior to starting the task, students need to have experience using a variety of resources to gather information. Go over the directions and answer any student questions. This problem requires the students to use a variety of resources to complete the project successfully. You need to judge how much help it is appropriate to give the students. Go over the directions and answer any student questions. This project is an appropriate one to require a written progress report about halfway through the project. When the projects are complete, have the groups give their oral presentations and collect the written reports.

Suggested Format: The students work in groups of three and are given a portion of three periods to work on the project. The oral reports and the processing will each require a portion of a class period. The project can be completed in six to ten school days.

Processing and Enrichment and Extension: See the previous problem.

(3) **"Ecologically Sound Community"**: Students determine a variety of ways to make their community more environmentally sound.

Sample student problem statement: The task for this problem is to determine ways to make your community more ecologically sound. The purpose of this task is to give you practice solving a non-routine problem affecting the environment and your life. Specifically:

1 For this problem, each group will research at least five (and a maximum of ten) ways to make this community more ecologically sound. You will probably need to use a variety of resources, including written and people resources.

2 Each group will prepare an oral presentation of two minutes or less on the gathered information. I will select the member of your group to present the oral report.

3 Your group will turn in a written report including a final list of ways to make this community more ecologically sound (minimum of five ways, maximum of ten) and a description of how you arrived at your final list and what resources you used, including resources that didn't result in any new ideas.

Teaching suggestions: Similar to the previous two problems, except the focus is on the students' school. In my experience, the first four problems in this strand form a logical sequence moving the students' focus from their individual situation, to their class, to the whole school, and to the local community.

(4) **"Environmentally Sound Products"**: Students develop a list of ten products that they believe would have the most positive effect on the environment if made available to people in the community.

Sample student problem statement: The task for this problem is to develop a list of ten products that would have the most positive effect on the environment if made available to the people of your community. The task gives you practice solving a nonroutine problem affecting the environment and your life. Specifically:

1   For this problem, each group will research at least ten products that would have a positive effect on the environment if made available to the people of your community. You will probably need to use a variety of resources, including written and people resources.

2   You want your products to be ecologically sound (minimize negative effects on the environment and maximize positive effects on the environment), but also ones that people would buy.

3   Each group will prepare an oral presentation of two minutes or less on the gathered information. I will select the member of your group to present the oral report.

4   Your group will turn in a written report including a final list of your ten products, a description of how you arrived at your final list, and what resources you used, including resources that did not result in any new ideas.

Teaching suggestions: Similar to the previous three problems, except the focus is on the students' local community. In my experience, the first four problems in this strand form a logical sequence moving the students' focus from their individual situation, to their class, to the whole school, and to the local community.

Suggested Format. The students work in groups of three and are given a portion of three periods to work on the project. The oral reports and the processing will each require a portion of a class period. The project can be completed in six to ten school days.

(5) **"Comparing Products":** Students compare the effect on the environment of certain products (e.g., cloth diapers versus disposable diapers).

Sample student problem statement: The task for this problem is to compare two similar products or situations for their effect on the environment. Not every group will be comparing the same items. The purpose of this task is to give you practice solving a nonroutine problem requiring the use of outside resources and involving an application affecting the environment. Specifically:

1   I will give each group two products or situations to compare concerning their effect on the environment. You will probably want to use both written and people resources to solve this problem.

2   Each group will prepare an oral presentation of two minutes or less on the gathered information. I will select the member of your group to present the oral report.

3   Your group will turn in a description of the process you used to evaluate the two items and your conclusions. You should

identify the written and people resources you used for data, including sources you tried but did not lead to useful data.

Teaching suggestions: Prior to starting the task, students need to have some familiarity with getting data from a variety of resources such as the library, magazines, internet, nonprofit organizations and other people. Go over the directions and answer any student questions. Each group will need items to compare. My suggestion is that you restrict it to two comparisons such as: comparing brown shopping bags to plastic shopping bags and comparing disposable diapers to reusable cloth diapers. I suggest these two because there are strong arguments for both sides of each issue and there is much written about both issues. If you prefer each group making their own comparison, then some other comparisons include electric stove versus a gas stove; a hamburger from beef locally grown versus a banana grown in Brazil; four people commuting to work in one car versus four people commuting to work by train; a weekend camping trip 300 miles away versus an overnight bed and breakfast vacation 50 miles away; and a cross-country telephone call versus a cross-country letter. Of course, comparisons of items more relevant to your local community or students are even better. For example, your class could brainstorm potential comparisons, with some of the above examples included to stimulate ideas, and then select from that list.

The students probably need some help in researching this problem. It might be helpful to brainstorm as a class or in the groups what types of issues need to be considered and what types of resources need to be explored. This problem is complex enough that it seems important to have them submit a progress report to you after the first week.

You need to be sure that students will have access to the resources they need, especially if you are including school library materials. For example, you may want to require students to not check out certain resources from the library so that all groups will have access to them. Perhaps you can arrange for certain articles to be copied for groups. It certainly would be appropriate to discuss this issue with the students. When the projects are complete, have the groups give their oral presentations and collect the written reports.

Evaluation and Processing. It seems essential in the processing of this project to focus on the use of resources to make the comparisons. One way to do that is to select good uses of resources from various groups and discuss these with the class. Also, you may ask each group to identify one time they felt they could not locate a resource they wanted, then see if as a class you can locate appropriate resources. Issues that involve effective examples of persistence also provide a good focus.

Suggested Format. The students work in groups of three and are given a portion of two periods to work before submitting the progress report, and a portion of two periods to complete the project and prepare their oral report. The oral reports and the processing will each require a portion of a class period. In my opinion this problem is worthy of a reasonable amount of time to give the students a chance to research the problem well. For many later problems, students will need to use a variety of resources; therefore, an investment of time on this problem can save time and improve the quality of work on later problems. Therefore, I suggest ten to fifteen school days for the project.

Enrichment and Extension. As a class, you can brainstorm additional items to investigate and then select a few to investigate. Trying to find items presently being compared in the school or the community would probably make the problem even more relevant.

(6) **"Ecologically Sound Lunch":** Students plan three lunches that are ecologically sound, relatively inexpensive, and enjoyable to eat.

Sample student problem statement: The task for this problem is to design an ecologically sound lunch. The purpose of this task is to give you practice solving a nonroutine problem in the field of ecology that requires you to use a variety of resources. Specifically:

1   You will plan a lunch including a beverage, main course, dessert and a "container" (e.g., lunch pail, brown bag, etc.). Your lunch will be evaluated primarily on its effect on the environment, but also on factors such as cost and taste.

2   You will turn in a written report including how you arrived at your final choice, what other lunches you considered (at least two others), and what resources you used.

3   Only a few students will be asked to orally explain their project, but each student should be prepared to give an oral presentation.

Teaching suggestions: Prior to starting the task, students need to have some familiarity with getting data from a variety of resources such as the library, magazines, internet, nonprofit organizations, and other people. Go over the directions and answer any student questions. You might want to emphasize that students will need to use a variety of resources to successfully complete the project.

Be sure students understand that their lunches will be evaluated on its effect on the environment, and on factors such as cost and taste. You might discuss how these types of considerations would be important in practical situations such as in marketing health food products. One option is to have students bring in their lunches and have lunch as a class. If not, there might be at least some discussion of whether students are willing to change their eating habits.

When the projects are complete, collect the written reports. If, in reviewing the reports, any oral presentations seem appropriate, it is probably best to schedule them before the general processing of the project.

Evaluation and Processing. Besides whatever comments naturally come out of reviewing all the reports, it seems important in the processing to focus on the types of resources used and how they were located. Obviously, any good examples of persistence in this project would be appropriate to discuss.

Suggested Format. The students work individually on this task. The oral reports, if any, and the processing will require a portion of a class period. The project can be completed in from six to ten school days.

Enrichment and Extension. One extension would be to investigate the school lunches and make recommendations for improvements.

(7) **"Environmentally Sound Fundraising":** Students determine a variety of ways to raise money for school groups that are environmentally sound.

Sample student problem statement: The task for this problem is to determine at least three methods to raise money for your school that are ecologically sound. The purpose of this task is to give you practice solving a nonroutine problem affecting the environment and your life. Specifically:

1 For this problem, each group will research at least three (and a maximum of ten) methods to raise money for your school that are ecologically sound. You will probably need to use a variety of resources, including written and people resources.

2 You want your methods to be effective at raising money and, also, ecologically sound (minimize negative effects on the environment and maximize positive effects on the environment).

3 Each group will prepare an oral presentation of two minutes or less on the gathered information. I will select the member of your group to present the oral report.

4   Your group will turn in a written report including a final list of methods to raise money for your school that are ecologically sound (minimum of three methods, maximum of ten) and a description of how you arrived at your final list and what resources you used, including resources that did not result in any new ideas.

Teaching suggestions: Prior to starting the task, students need experience using a variety of resources to gather information. Go over the directions and answer any student questions. You need to judge how much help it is appropriate to give the students. For example, many good ideas or leads can come from talking to appropriate people (or organizations) in the community. Do you think that at least some of your students would use people resources without your guidance? If not, you may want to suggest to students that they brainstorm in their group at least a few people they might talk to for ideas. In any case, this would be a good problem for students to start by brainstorming possible ideas. This project is an appropriate one to require a written progress report about halfway through the project.

When the projects are complete, have the groups give their oral presentations and collect the written reports. As in the previous problems in this strand, a good focus for the processing is what resources people used to complete the project. Selecting and discussing a variety of resources students used is a good way to focus the discussion, especially if the discovery of a resource involved a good example of persistence. As part of the project, your class might select a few of the best ideas and use them for fundraising. Also, your class might try to encourage other groups to use your fundraising ideas.

Suggested Format. The students work in groups of three and are given a portion of three periods to work on the project. The oral reports and the processing will each require a portion of a class period. The project can be completed in six to ten school days.

Enrichment and Extension. The class can compile the suggestions, perhaps research additional methods, and make a handout, booklet or internet resource that would be made available to interested people in the school and/or community. Of course, if the results can be integrated into another nonroutine problem such as "Planning a trip," that would facilitate the students appreciating the power of the curriculum.

There is one problem at the advanced level for the ecology strand, **"Profit and the Environment":** Students determine a method for calculating profit that takes into account the effect on the environment.

Sample student problem statement: The task for this problem is to determine as a group how to determine profit in a way consistent with principles of ecology. The purpose of this task is to give you practice solving a nonroutine problem involving environmental issues and developing a procedure. Specifically:

1   Your group has been asked to develop a method for evaluating businesses by an environmental group. One method for evaluating businesses is by determining profit as calculated by the difference between income and expenditures. The environmental group wants you also to consider the effect on the environment of the business as well as other factors that might affect the quality of life of the employees and others affected by the business.

2   You are to develop a written report including what factors your group considers important and why, including profit (you can redefine profit), environmental factors, and at least two other categories; as well as how you would evaluate these factors and the business in general. This could be a formula or a procedure to follow that would result in a rating of the company (e.g., excellent, very good, fair, poor); and a demonstration of your method on two real businesses.

3   Each group will prepare an oral presentation of two minutes or less on the gathered information. I will select the member of your group to present the oral report.

Teaching suggestions: This is a very meaningful nonroutine problem that investigates how our use of the traditional formulas to determine profit can directly or indirectly affect the quality of life of people affected by the company's actions. However, this problem is certainly advanced in that it requires students to look critically at the concept of profit that has basically been unchanged for many years in terms of its emphasis on the difference in money needed to make a product and money collected selling the product, the formula ignoring other factors such as effect of the company's actions on the environment. Of course, with much more sophisticated formulas to calculate that profit developed over the years. Therefore, it is appropriate for most classes to have some introduction to the issue before working on the problem. For example, you and the class can review an article(s) related to the problem. One source might be an article on Bhutan's emphasis on replacing the concept of GNP (Gross National Product) with GNH (Gross National Happiness) and the work of a variety of economists to develop scientific approaches to determining GNH.

Go over the directions and answer any student questions. You might want to point out that their methods will be somewhat subjective. The factors they decide are important (other than environmental) could differ from another group's due to different values and beliefs. However, you will be able to evaluate the quality of the group's process of arriving at its values and whether the method well measures their stated values. You might want to limit the businesses to out of state businesses, or at least not in the vicinity of the school, to avoid difficult situations for students who have parents working in the businesses. In the written report, it should be clear to students that in explaining why they chose their criteria, they should comment on other criteria they considered and why they were not used. If neither the option of using local businesses nor out-of-town businesses feels comfortable for your context, then it might be appropriate to not require the evaluation of any actual company. This is a good project to require a progress report halfway through the project.

When the projects are complete, have the groups give their oral presentations and collect the written reports.

Evaluation and Processing. Besides whatever comments naturally come out of reviewing all the reports, it seems important in the processing to evaluate whether the project resulted in them viewing businesses differently. If yes, that experience might serve to illustrate how the quality of one's life in terms of one's values and views can be positively affected by the process of solving nonroutine problems. Hopefully, this is a concept students can appreciate by this point in the curriculum.

Suggested Format. The students work in groups of three and are given a portion of three periods to work on the project and prepare their oral report. The oral reports and the processing will each require a portion of a class period. The project can be completed in ten to fifteen school days.

Enrichment and Extension. The class can try to come up with one procedure for evaluating businesses that most reflects the values of the class. Of course, that procedure could then be used to evaluate businesses, either local or not.

There are four problems in the strand that focuses on appreciating diversity:

(1) **"Interesting Rituals and Customs":** Students find at least five interesting rituals or customs practiced in their community.

Sample student problem statement: The task for this problem is to discover some interesting rituals and customs practiced in your community. The purpose of this task is to solve a problem which might increase your appreciation for the diversity of people and cultures. Specifically:

1   Your group will find between five and ten customs or rituals practiced in your community. For this project, any ritual or custom practiced over at least the last three years by at least one person in the community will be acceptable. The ritual or custom could be associated with a religion, an ethnic group, or a group of friends or acquaintances. Two examples of an acceptable custom or ritual not associated with an ethnic group are, first, if a group of people meet each year for the spring solstice and take a hike to celebrate the coming of spring and, second, if members of a family get together once a year, cook a certain type of meal and then tell family stories.

2   You will split your list into two groups, rituals or customs that this class would enjoy doing as a class, and rituals or customs that are interesting, but might not be ones that the class would want to try.

3   Each group will prepare an oral presentation of two minutes or less on the gathered information. I will select the member of your group to present the oral report.

4   Your group will turn in your list of customs and rituals and a written description of the process you used to come up with your list.

Teaching suggestions: Go over the directions and answer any student questions. Students may need some additional explanation to understand what they are looking for. In that case, you may want to add some more examples of acceptable rituals and customs. Also, you can allow groups to check with you if they are not sure whether a certain practice qualifies or not. Students should realize that the major way to gather good examples is by talking to a variety of people to unearth customs and rituals in the community. Groups may want to start by brainstorming people or groups that would be appropriate to talk to or interview. If needed, you may want to discuss and/or roleplay such a conversation to insure that students are sensitive to how to collect such data appropriately. When the projects are complete, have the groups give their oral presentations and collect the written reports.

Processing. A good focus for the processing might be a discussion of how talking to people can provide excellent data for solving problems. Students could be asked, "Were you surprised by what you found out from some of the people you questioned? Did you think you would get as many interesting customs and rituals as you got?" Since one purpose of this problem is to increase the student's appreciation of diversity, including appreciation of a variety of ethnic and religious groups, you may want to ask students if they learned anything new about other cultures or religious groups. For closure, it could be interesting to pick at least one of the customs to do/observe together as a class.

Suggested Format. The students work in groups of three and are given a portion of two periods to work on the problem. The oral reports and the processing will each require a portion of a class period. The project can be completed in five to ten school days.

Enrichment and Extension. Each group can research interesting customs and rituals outside the community, using primarily written resources.

(2) **"Planning a Cultural Trip":** Students plan a class trip that is inexpensive, enjoyable and increases the students' appreciation of the diversity of cultures.

Sample student problem statement: The task for this problem is to plan a trip that will increase your appreciation of other cultures. The purpose of this task is to solve a problem that requires you to use written and people resources that will increase your appreciation of other cultures. Specifically:

1   Each group is to plan a one-day class trip. The trip should have three characteristics: it has educational value and will increase your appreciation of other cultures; it is interesting and enjoyable from the point of view of students and it is inexpensive. Your group's plan should include transportation, food considerations, schedule, costs and a description of the day's activities.

2   The first activity in your groups will be to brainstorm what resources your group can use to plan this trip, including written and people resources. In addition, your group can prepare a questionnaire to gather data from the students concerning their interests in various ideas.

3   Each group will prepare an oral presentation of two minutes or less on the gathered information. In your presentation, you are trying to convince the class that your group's proposed trip is

the best. I will select the member of your group to present the oral report.

4 Your group will turn in a written report including what resources you used, how you decided on the details of the trip and a description of the trip.

Teaching suggestions: This project is fairly complex and requires students to find and use a variety of resources. You need to judge whether the students are experienced enough with using resources that they can complete the project successfully. It may be appropriate for the class to work on parts of the project (or even the entire project) as a class with your guidance. For example, you might want to have the students plan the activities and then as a class investigate transportation and details such as where to eat. Another option is to include a fundraiser as part of the project; this can be incorporated into the group work or done as a class.

Go over the directions and answer any student questions. You might want to emphasize the need to brainstorm and investigate a variety of resources in order to successfully complete the project. The project is best when students know that the class will actually take one of the trips; this helps them see directly the results of their problem solving activities. You might want to require at least one questionnaire from each group to gather feedback from students concerning their interest in various ideas. You might want to have the questionnaires all handed out the same time and remind students that ideas on the questionnaires cannot be used by another group.

This is a good project to require a written progress report approximately halfway through the project. When the projects are complete, have the groups give their oral presentations and collect the written reports.

Suggested Format. The students work in groups of three and are given a portion of two periods to start the project and develop their questionnaires, and a portion of two periods to analyze the data and complete the project. The oral reports and the processing will each require a portion of a class period. The project can be completed in ten to fifteen school days. This is a project that it is worth extending the time allotted to insure a good final product.

Processing. Besides whatever comments naturally come out of reviewing all the reports, it seems important in the processing to discuss how students used resources to solve this problem, including examples of false leads. If the class actually takes the trip, a discussion of how the trip turned out compared to how it was planned could be interesting. In my experience, this is an excellent nonroutine problem for the students to concretely see how the steps can result in an excellent result. For example, for class trips to New York City, my students consistently enjoyed an educational trip significantly more fun and less expensive than the typical school trip.

Enrichment and Extension. If time permits, a second similar trip or planning an individual one-day or weekend trip would certainly be an excellent extension. I have included below a description of one such class trip planned by my problem solving class, as well as some thoughts concerning the relationship to mathematics education:

Problem solving trip: Boston

In the field-testing of the curriculum, one group planned a trip to Boston. Their research uncovered many good ideas including (a) a variety of relevant museums such as the Museum of the National Center of Afro-American Artists, Cambridge Multicultural Arts Center and the African Meeting House; (b) relevant sights in culturally rich sections of Boston; (c) Cultural Survival, an organization that focuses on helping indigenous cultures survive and (d) interviewing Napoleon Jones Henderson, an Afrocentric artist. Perhaps the most interesting idea was to plan an afternoon with a group of inner city students trained to be peer leaders for an after school program. The idea was to share skills, we would lead a problem solving session and they would lead a community building activity.

How is this problem connected with the mathematics curriculum? Firstly, the students see how the three steps of solving nonroutine problems are useful in solving this practical problem. Specifically, students who have successfully solved a sequence of nonroutine problems might demonstrate the following type of evidence of applying the three steps:

1 Problem recognition and orientation: Students might realize that some of the criteria for evaluating their project are contradictory, and part of a good solution will be finding possibilities that satisfy all three criteria. For example, many possibilities for the trip that would satisfy the criteria of being interesting or educational might be expensive. In addition, given the criteria of developing a trip appealing to their classmates (versus their small group only), they might realize the need to gather data from classmates (or other peers) rather than relying solely on their own ideas.

2 Trying something: Students might brainstorm possible sources for ideas (e.g., people to talk to, written sources to check) and

decide which resources to actually investigate. They might decide to develop a questionnaire for their fellow students (or similar students such as students in another class) to answer in order to gather data.

3   Persistence: Students might realize that there is a need for additional data to finish the problem well. For example, most questionnaires give some clear data, but also generate some data that suggests the need to ask follow-up questions. Or students might realize a need to persist by gathering additional information in some key areas suggested by their initial data.

In addition, even in a problem that does not seem to be "mathematical" in nature, students many times improve in specific mathematical content areas. For example, in "Planning a Trip," after the groups report out concerning their ideas, there might be ten suggestions for portions of the trip. The students can be given the task of developing a method to quickly, yet accurately, evaluate which of the ideas are most liked by the class. After a few problems of this type, the students get a sense of how to quantitatively generate the right amount of data to answer the question quickly, yet accurately. For example, students might compare three methods: each student voting for one idea they most like, each student voting for three ideas they most like and each student rank ordering all ten ideas. Students discover that one vote per student is typically not enough data to draw comprehensive conclusions, and that rank ordering all the items is too much data in the sense that it takes significantly longer to evaluate the data than three votes per student, while basically generating the same quality of results. In addition, students also get a sense of when additional data is needed to clarify the results. For example, after voting there might be one group of three ideas that are clearly preferred by the class and a second group of several additional ideas that are basically statistically equivalent in terms of the students' preference, while being clearly less preferred than the first group and more preferred than the remaining ideas. Students develop a sense that a quick additional vote on the second group can usually clarify which of the items in this group are most preferred. Finally, students begin to appreciate that while this type of quantitative analysis gives data quickly that is valuable in making a decision, there is typically still a need for additional discussion to make a decision that "works." For example, given time and cost limitations which combination of ideas makes sense?

Sample edited (shorten) permission slip: The Problem Solving class is planning a trip to Boston on Tuesday, May 23 approximately 8:30 am to 7:00 pm, Wednesday, May 24. The trip is the result of a student project to plan an inexpensive, enjoyable and educational trip to increase appreciate of diversity. The trip will include a joint workshop with an inner city program for training peer leaders, visit to the North End, Cambridge, and Copley Square and a museum or exhibit featuring an appreciation of diversity. The cost of the trip will be $25 for van transportation and parking (if enough students go, this cost will be reduced). In addition, students will need money for expenses (e.g., food). Students will be on their own for a substantial portion of the trip. For example, the students will have approximately two hours in the Copley Square area. They will be restricted to a specified area for which they will have a map. During these times the students will have directions, information and parameters which, if followed, will insure that they have an enjoyable time and return to the designated meeting places when required. There will be two chaperones on the trip.

(3) **"Finding a speaker":** Students find an inexpensive, interesting speaker on the topic of improving appreciation of cultural diversity.

Sample student problem statement: The task for this problem is to find a speaker that will increase the class's appreciation of the diversity of ethnic groups or cultures. The purpose of this task is to solve a problem that requires you to use written and people resources and will increase the class's appreciation of the diversity of ethnic groups and cultures. Specifically:

1 Each group is to find three speakers, who would be available to speak to this class, with three characteristics: his or her presentation would have educational value and would increase the class's appreciation of other ethnic groups and cultures; his or her presentation would be interesting and enjoyable from the point of view of students; and his or her presentation would be free or inexpensive.

2 The first activity in your groups will be to brainstorm what resources your group can use to find the speakers, including written and people resources. In addition, your group can prepare a questionnaire to gather data from the students concerning their interests in various ideas.

3 Each group will prepare an oral presentation of two minutes or less on the gathered information. In your presentation, you are trying to convince the class that one of your group's speakers is

the best. I will select the member of your group to present the oral report.

4 Your group will turn in a written report including what resources you used, how you decided on the three speakers and a description of each speaker's presentation.

Teaching suggestions: Like the previous problem, this project is fairly complex and requires students to find and use a variety of resources. You need to judge whether your students are experienced enough with using resources that they can complete the project successfully. It may be appropriate for the class to work on parts of the project (or even the entire project) as a class with your guidance. Your goal is to make the problem challenging but not frustrating or too straightforward.

Go over the directions and answer any student questions. You might want to emphasize the need to brainstorm and investigate a variety of resources in order to successfully complete the project. The project is best when students know that the class will invite at least one of the speakers to their class; this helps them see directly the results of their problem solving activities. You might want to require at least one questionnaire from each group to gather feedback from students concerning their interest in various ideas. You might want to have the questionnaires all handed out the same time and remind students that ideas on the questionnaires cannot be used by another group. This is a good project to require a written progress report approximately halfway through the project. When the projects are complete, have the groups give their oral presentations and collect the written reports.

Processing: Besides whatever comments naturally come out of reviewing all the reports, discuss how students used resources to solve this problem, including examples of false leads. If the class actually invites a speaker, a discussion of how the speaker turned out compared to how it was planned could be interesting.

Suggested Format. The students work in groups of three and are given a portion of two periods to start the project and develop their questionnaires, and a portion of two periods to analyze the data and complete the project. The oral reports and the processing will each require a portion of a class period. The project can be completed in ten to fifteen school days. This is a project that it is worth extending the time allotted to insure a good final product.

Enrichment and Extension. The students can plan a second speaker (perhaps with a different focus) later in the year. A more ambitious project would be to arrange for a speaker for the whole school and/or the local community.

(4) **"Increasing Diversity":** Students determine how an environmental group can increase the diversity of its membership and better serve a more diverse group of people.

Sample student problem statement: The task for this problem is to make recommendations to a nonprofit environmental group that wants to increase the diversity of its members. The purpose of this task is to solve a problem that requires you to use written and people resources and will increase the class's appreciation of the diversity of ethnic groups and cultures. Specifically:

1   For this task, assume your group has been asked by a nonprofit environmental group to help them increase the diversity of their members. Presently, their membership is primarily white with average or above average income. The group not only wants to increase the numbers of members from other groups, but also better address environmental issues that affect a broader group of people. The group is willing to invest a maximum of $1,000 to implement your suggestions.

2   Each group is to make five to ten recommendations to the group. The first activity in your groups will be to brainstorm what resources your group can use to complete this task, including written and people resources.

3   Each group will prepare an oral presentation of two minutes or less on the gathered information. In your presentation, you are trying to convince the class that your recommendations will be most effective. I will select the member of your group to present the oral report.

4   Your group will turn in a written report including what resources you used, how you decided on the recommendations and a description of each recommendation.

Teaching suggestions: As in the other problems in this strand, this nonroutine problem is fairly complex and requires students to find and use a variety of resources. You need to judge whether the students are experienced enough with using resources that they can complete the project successfully. It may be appropriate for the class to work on parts of the project (or even the entire project) as a class with your guidance.

Go over the directions and answer any student questions. You might want to emphasize the need to investigate a variety of resources in order to successfully complete the project. Be sure that students understand what the nonprofit group wants, especially that they are limited in funds ($1,000) and that they want not only to increase the numbers of members from other groups, but also to better address environmental issues that affect a broader group of people. You might mention to the class that the presented problem is similar to actual problems a variety of organizations have attempted to address in the past.

This is a good project to require a written progress report approximately halfway through the project. When the projects are complete, have the groups give their oral presentations and collect the written reports.

Processing. Besides whatever comments naturally come out of reviewing all the reports, it seems important to discuss how students used resources to solve this problem, including examples of false leads and good persistence. It would be appropriate to ask students if, as a result of the project, they better appreciate or understand a variety of cultural and ethnic groups.

Suggested Format. The students work in groups of three and are given a portion of three periods to complete the project. The oral reports and the processing will each require a portion of a class period. The project can be completed in ten to fifteen school days.

Enrichment and Extension. A good extension for this project would be to invite an expert on this issue to speak to the class and react to their research. In addition, the class could decide on the ten best recommendations and see if there are any local (or national) environmental groups interested in the results of the project.

9

Practical applications affecting students' lives

Introduction

The conceptual framework and pedagogy indicate the need for nonroutine problems in the curriculum that emerge from the unique context of the individual students, as well as the class, school and local community. One purpose of these problems is to increase the meaningfulness of the curriculum and the likelihood that the students will transfer the skills learned in the curriculum to their individual lives. Chapter 7 addressed nonroutine problems that concern establishing a supportive learning environment, as well as experience with basic researching tools needed in later problems. Chapter 8 addressed what we can label as nonroutine problems concerning non-mathematical content in two specific strands of content, "Increasing appreciation of diversity" and "Ecological awareness and sustainability." This chapter provides specific guidelines and examples of how to develop this component of the curriculum in the reader's specific unique context. These are nonroutine problems that are essential to the curriculum; that is, problems that are clearly relevant and meaningful to the students in your specific classroom and context. From my experience, I believe these problems are essential to help insure transfer to the students' lives, meaning their ability to recognize meaningful nonroutine problems in their lives and address those problems effectively. For example, in my one-year problem solving class, students were required to identify and address at least two nonroutine problems in their life. As I will discuss later in this chapter, I believe this requirement and their work to satisfy the requirement not only

DOI: 10.4324/9781003393283-9

were necessary to insure transfer, but also provided the clearest evidence of the effectiveness of the curriculum! As implied in the chapter on pedagogy, what is essential is that you, the teacher, need to be sensitive to "naturally arising" non-routine problems that are meaningful to your students. The discussed problems in these chapters are only examples that hopefully provide you with the foundation to identify the nonroutine problems most relevant to your context.

Specifically, this chapter explores three types of problems: (1) nonroutine problems that focus on problems directly related to the students' classroom, school and/or community, (2) individual and class nonroutine problems specific to their context and (3) a few additional nonroutine problems that tend to be meaningful to most students. In addition, the last section of this chapter briefly discusses some methods of assessing whether the students can transfer their ability to solve nonroutine problems to solving meaningful (non-mathematical) nonroutine problems in their life.

Nonroutine problems directly related to the students' classroom, school and/or community

The three problems in this section are all similar and require the students to improve their classroom, school and local community, generating data specific to their context:

(1) **"Improving the Class":** Students generate a list of ways to improve the class based on collected data.

Sample student problem statement: The task for this problem is to identify ways that this class can be improved. The purpose of this task is to give you practice solving problems using a questionnaire as a tool and requiring persistence. Specifically:

1 Each group will prepare a list of between five and ten ways to improve this class. Your work on this task will be in two parts. In the first part, your group will brainstorm ideas for improving the class and, based on the ideas, prepare a list of tentative ideas for improving the class and a questionnaire to gather information from students in this class to determine whether your ideas are good. I will copy and distribute the questionnaires and return the completed questionnaires to your group.

2 In the second part, you will analyze the data from the questionnaires and prepare your final list. To prepare your final list, your group may construct and administer a second questionnaire as

long as the questionnaire addresses items on your first list or new items not on another group's original list. Your group's final list can only contain ideas on your original list or new ideas not on another group's original list.

3 Each group will hand in a written report including the group's final list of improvements and documentation of the group's process of selecting the final list. Each group will prepare an oral presentation of two minutes or less on your work.

Teaching suggestions: This problem can be introduced by telling students that this new problem they are going to work on will directly affect their life and, hopefully, illustrate to each student the relevance of problem solving in their life. Go over the directions and try to answer any questions. Be sure that students understand the two stages of completing the project. It might help to indicate to students the need for a questionnaire; that is, the questionnaire will provide data to judge whether the ideas that the group believes are good are, in fact, good in the opinion of other students. Citing examples, such as screenings for potential ads and polls on certain issues, might clarify the need for this step.

Be sure that students realize that their first list of ideas is not restricted to items generated in the brainstorming; e.g., there will be time between the brainstorming and the completion of the first list in which they can generate additional ideas to add to their list.

This problem is particularly powerful if the students know that you are committed to making at least some of the changes suggested. You might indicate to students that they may want to check with you whether certain ideas are acceptable. You might emphasize that the problem is to come up with ideas that are good in the eyes of both the teacher and the students, that is what makes it a good problem!

Students may not appreciate the purpose of the second questionnaire to clarify unclear results of the first questionnaire and to gather information about ideas raised by the data from the first questionnaire. The second questionnaire is connected with the need for persistence in problem solving. Many students do not perceive a need for persistence. You may want to either alert them to this issue during the project or wait until the processing to discuss whether groups chose to give a second questionnaire and how that decision affected the quality of their results.

Comment. It seems important that the students' work actually results in tangible changes in the class. Indeed, assuming teacher (and student) willingness, this problem solving process should result in clear improvements in the class. In fact, if no good ideas result from the project, the students could not have been carrying out the steps correctly and it would be worthwhile to retry the process.

When the projects are complete, have the groups give their oral presentations and collect the written reports.

Processing. A major focus of the processing should be the role of the two questionnaires in solving the problem. Discussion of examples of how the data from the questionnaires influenced the final lists would be fruitful, particularly examples in which the data was counter to the intuition of a group. In addition, a discussion of how the problem solving methods facilitated the improvement of the class would be appropriate.

Suggested Format. The students work in groups of three and are given a portion of two periods for the first part and an additional portion of two periods for the second part. The processing and oral presentations should each require a portion of a class period and perhaps a second class to discuss implementing some of the suggestions. The project can be completed in ten to fifteen school days.

Enrichment and Extension: Students can use the same methods to identify ways of improving the high school in general. Or the students can repeat the project a second time. This option would probably give the students a good sense of how persistence can produce a better quality of solutions.

(2) **"Improving the School"**: Students prepare a list of five to ten ways to improve the school. They develop questionnaires to help them with the project.

Sample student problem statement: The task for this problem is to identify ways that this school can be improved. The purpose of this task is to give you practice solving problems using a questionnaire as a tool and requiring persistence. Specifically:

1 Each group will prepare a list of between five and ten ways to improve this school. Your work on this task will be in two parts. In the first part, your group will brainstorm ideas for improving the school and, based on those ideas, prepare a list of tentative ideas for improving the school and a questionnaire to gather information from students in this class to determine whether your ideas are good. I will copy and distribute the questionnaires and return the completed questionnaires to your group.

2 In the second part, you will analyze the data from the questionnaires and prepare your final list. To prepare your final list, your group may construct and administer a second questionnaire as

long as the questionnaire addresses items on your first list or new items not on another group's original list. Your group's final list can only contain ideas on your original list or new ideas not on any other group's original list.

3   Each group will hand in a written report including the group's final list of improvements and documentation of the group's process of selecting the final list. Each group will prepare an oral presentation of two minutes or less on your work.

Teaching suggestions: This problem can be introduced by telling students that this new problem they are going to work on will directly affect their life and, hopefully, illustrate to each student the relevance of problem solving in their life. Go over the directions and try to answer any questions. Be sure that students understand the two stages of completing the project. It might help to indicate to students the need for a questionnaire; that is, the questionnaire will provide data to judge whether the ideas that the group believes are good are, in fact, good in the opinion of other students. Citing examples, such as screenings for potential ads and polls on certain issues, might clarify the need for this step. This problem is particularly powerful if the students know that you are committed to helping them initiate at least some of the changes suggested. You might indicate to students that they may want to check with you whether certain ideas are acceptable. You might emphasize that the problem is to come up with ideas that are good in the eyes of both the teacher and the students.

Students may not appreciate the purpose of the second questionnaire, to clarify unclear results of the first questionnaire and to gather information about ideas raised by the data from the first questionnaire. The second questionnaire is connected with the need for persistence in problem solving. Many students do not perceive a need for persistence. You may want to either alert them to this issue during the project or wait until the processing to discuss whether groups chose to give a second questionnaire and how that decision affected the quality of their results.

Comment. It seems important that the students' work actually results in tangible changes in the school. Indeed, this problem solving process should result in clear improvements in the school. If no good ideas result from the project, the students could not have been carrying out the steps correctly and it would be worthwhile to retry the process. When the projects are complete, have the groups give their oral presentations and collect the written reports.

Processing. A major focus of the processing should be the role of the two questionnaires in solving the problem. Discussion of examples of how the data from the questionnaires influenced the final lists would be fruitful, particularly examples in which the data was counter to the intuition of the groups. In addition, a discussion of how the problem solving methods facilitated the improvement of the school would be appropriate.

Suggested Format. The students work in groups of three and are given a portion of two periods for the first part and an additional portion of two periods for the second part. The processing and oral presentations should each require a portion of a class period, and perhaps a second class to discuss implementing some of the suggestions. The project can be completed in ten to fifteen school days.

Enrichment and Extension. This is a project that can be repeated a second time, trying to use some of the suggestions from the processing to find additional ways to improve the school. From an instructional point of view, a second attempt would present an excellent opportunity to fine tune some of the skills required in this task.

(3) **"Improving the Community":** Students determine actions which could improve the quality of the community in which they live.

Sample student problem statement: The task for this problem is to identify ways that the community you live in can be improved. The purpose of this task is to give you practice solving problems using a questionnaire as a tool and requiring persistence. Specifically:

1   Each group will prepare a list of between five and ten ways to improve this community from the point of view of the students in this class. Your work on this task will be in two parts. In the first part, your group will brainstorm ideas for improving the school and, based on those ideas, prepare a list of tentative ideas for improving the community and a questionnaire to gather information from students in this class to determine whether your ideas are good. I will copy and distribute the questionnaires and return the completed questionnaires to your group.

2   In the second part, you will analyze the data from the questionnaires and prepare your final list. To prepare your final list, your group may construct and administer a second questionnaire as long as the questionnaire addresses items on your first list or new items not on other group's original list. Your group's final list can only contain ideas on your original list or new ideas not on any other group's original list.

3   Each group will hand in a written report including the group's

final list of improvements and documentation of the group's process of selecting the final list. Each group will prepare an oral presentation of two minutes or less on your work.

Teaching suggestions: This problem can be introduced by telling students that this new problem they are going to work on will directly affect their life and, hopefully, illustrate to each student the relevance of problem solving in their life. Go over the directions and try to answer any questions. Be sure that students understand the two stages of completing the project. By this point in the curriculum students should appreciate the need for a questionnaire and the purpose of a second questionnaire if appropriate, to clarify unclear results of the first questionnaire and to gather information about ideas raised by the data from the first questionnaire. If this is not the case, then you should review the need.

This problem is particularly powerful if the students know that you are committed to helping them initiate at least some of the changes suggested. You might indicate to students that they may want to check with you whether certain ideas are acceptable. You might emphasize that the problem is to come up with ideas that are good in the eyes of both the teacher and the students, that is what makes it a good problem! When the projects are complete, have the groups give their oral presentations and collect the written reports.

Comment. It seems important that the students' work actually results in tangible changes in the community. Indeed, this problem solving process should result in clear improvements in the community.

Processing. A major focus of the processing should be the role of the two questionnaires in solving the problem. Discussion of examples of how the data from the questionnaires influenced the final lists could be fruitful, particularly examples in which the data was counter to the intuition of the groups. In addition, a discussion of how the problem solving methods facilitated the improvement of the community would be appropriate.

Suggested Format. The students work in groups of three and are given a portion of two periods for the first part and an additional portion of two periods for the second part. The processing and oral presentations should each require a portion of a class period and, perhaps a second class to discuss implementing some of the suggestions. The project can be completed in ten to fifteen school days.

Enrichment and Extension. This is a project that can be repeated a second time, trying to use some of the suggestions from the processing to find additional ways to improve the community. From an instructional point of view, a second attempt would present an excellent opportunity to fine tune some of the skills required in this task.

Individual and class nonroutine problems meaningful in your context

As implied previously, a major goal of the curriculum is that the students can apply the skills involved in solving nonroutine problems in mathematics to solving significant nonroutine problems in their lives. This section is divided into two parts, the first discusses individual meaningful nonroutine problems that students define and solve in their personal life. The second part discusses meaningful nonroutine problems defined and solved by your class.

Individual problems

In my classes, after considerable experience solving nonroutine problems, students are required to define and solve meaningful nonroutine problems in their lives. For example, in the second half of a full-year course in problem solving or in years three and four (two each year) in a four-year academic program. For individual problems, students typically select problems involving a major purchase (e.g., used automobile or a stereo system), obtaining employment (e.g., a well-paying, interesting summer job), planning a trip or vacation, budgeting (e.g., budgeting to buy an automobile) or selecting a post-secondary institution to attend. Two examples that indicate the potential of this approach are taken from two vocational students of poor mathematics ability who took my Problem Solving course. The first student had a lawn care business with four clients with large lawns. He had difficulty organizing the business and was ready to give it up. His problem was to effectively organize the business. To gather information, he interviewed his clients and three lawn care professionals. His initial conclusions included (a) plan a schedule to complete the lawns in four days, allowing for bad weather and other complications and (b) schedule your largest lawn early in the week. Based on these conclusions and other data, he devised a tentative plan that he checked with his clients and a fourth lawn care professional. He successfully implemented the plan and was very satisfied with the results.

The second student picked a very personal problem. He was concerned about his perceived inability to relate well with other students, particularly his inability to make close friends. He investigated a number of options and decided to see the school psychologist. I saw the psychologist the next fall after the student graduated and she indicated that he continued seeing her through the summer, making significant progress.

Below I include the sample student problem statement for the first such individual problem:

The task for this problem is to define and solve a problem which is significant to you. The purpose of this task is to give you practice solving nonroutine problems in your own life. Specifically:

(1) You will need to define your problem and have the problem approved by me. You will submit a written summary of the problem, how you intend to work on solving it, and how long you believe it will take you to complete the project. The problem will be approved if I am convinced that the problem is meaningful to you and the problem is significant (i.e., the problem would require the three steps to solve it).

(2) When the project is completed, you will turn in a written report including your solution to or progress for the problem, and a description of how you arrived at your solution. You should include anything you learned about problem solving in your description.

(3) Only a few students will be asked to orally explain their project, but each student should be prepared to give an oral presentation. If your project involves personal information you do not wish to share with the class, you will have the option of not sharing.

Teaching suggestions: This problem requires the student to recognize, define, and solve a problem relevant to him or her. This is quite a step from solving problems given by someone else. Therefore, you should expect that students might have difficulty. It could be useful to start the problem by having the classroom brainstorm possible problems (either personal or problems they imagine someone else might have) and then discussing the brainstorm. It might even be appropriate to return to the brainstorm after students have had time to think about the brainstorm (e.g., homework assignment).

Go over the directions and answer any student questions. I recommend that the deadline for completion of the projects be flexible, depending on the difficulty and nature of the problem. Perhaps a two or three week deadline is suggested with the understanding that, with permission, the deadline could be extended for appropriate individuals.

Since this problem is the first of four similar problems over two years, your emphasis should be on this first problem being a successful experience and a transition to a deeper understanding and appreciation of the usefulness of the steps in one's life. For example, some students might need direct help coming up with an idea. Others might require guidance based on your review of their proposed problem and plan. Some students might select too large a project and need help focusing the problem. Other students may require no guidance.

It would be appropriate to schedule progress reports for this project, perhaps more than one in some cases. As the projects are completed, collect the written reports. If, in reviewing the reports, any oral presentations seem appropriate, it is probably best to schedule them before the general processing of the project.

Processing. If a few projects require significantly more time than the others, you might want to schedule the processing when most of the projects are completed so as not to lose the freshness of the experience. Besides whatever comments naturally come out of reviewing all the reports, it seems important in the processing to discuss the experience of defining and solving an individual problem for the first time.

Suggested Format. The students work individually on this task. The oral reports, if any, and the processing will require a portion of a class period. The project can be completed in ten to fifteen school days, with the option of extensions for appropriate students. The option of allowing a pair of students to work on a problem meaningful to both of them, if suggested by a pair of students (versus as a stated option for the class), may be appropriate to allow. It may be appropriate for some individual students to extend their individual projects either as extra credit or instead of the next nonroutine problem.

Nonroutine problems generated and solved by your class

When students demonstrate a reasonable understanding of solving nonroutine problems, I require them to formulate and solve meaningful nonroutine problems in their own life. Typically, before assigning individual problems such as above, I have students identify and solve a significant nonroutine problem generated by the class. For example, one year a class of students in my Problem Solving class identified their primary significant problem as how to make the ceremony of graduation meaningful for them. Most of the students were seniors barely meeting the requirements for graduation and not students that typically would be individually recognized at graduation. The class solved two problems related to graduation. First, they complained that graduation was planned to be inside the school rather than an outdoors ceremony as in some past years. They were able to work with the school leaders (with whom they usually did not associate) and successfully convince the administration to hold the ceremony outdoors. Second, we decided to complete the final exam for the class during the last week of classes, allowing the two-hour scheduled period for our final to be used for a graduation ceremony that was both meaningful and enjoyable. They decided that each student could invite two guests and that each student would have to prepare a five-minute "presentation" to talk about the meaning of the day for them and they determined a good location outside the school building (i.e., a fieldtrip with permission slips, etc.). In addition, the students determined, using mathematics as a tool, what foods and music to include in the ceremony to best please all the class members. The ceremony was quite meaningful and probably will be remembered by the students for a long time! I must comment that whenever I work with a class to define and solve a meaningful nonroutine problem, I am always amazed at the quality of the solutions we are able to generate as a group, which are always clearly better than any solution I or any student alone could generate.

Here is a sample student statement for generating nonroutine problems:

The task for this problem is to identify at least three nonroutine problems that the students in this class believe are meaningful for the class to solve or address. The purpose of this task is to give you practice solving a meaningful nonroutine problem involving the use of a questionnaire. Specifically:

(1) Each group will identify at least three nonroutine problems that the students will agree are meaningful to be addressed by the class. In addition, you should pick problems valuable from the point of view of improving your ability to solve nonroutine problems. It is suggested that each group start by brainstorming ideas for the project.

(2) Your group will be able to develop and give out a questionnaire that can be answered in ten minutes or less to gather information. Any nonroutine problem not mentioned on your questionnaire and mentioned on another group's questionnaire cannot be used on your final list.

(3) Each group will prepare an oral presentation of five minutes or less to convince the class that your nonroutine problems would be the most meaningful and interesting to study. I will select the member of your group to present the oral report.

(4) Your group will turn in a written report including a list of the three nonroutine problems and a description of the process your group used to arrive at your list.

Teaching suggestions: Go over the directions and answer any student questions. Indicate to students that they can ask you for your judgment concerning the merit of the problem from the point of view of improving their ability to solve nonroutine problems. You might want to indicate to the students that they are primarily looking for problems that will interest other students. Your evaluation will only be whether it is truly a nonroutine problem and will not be affected by whether mathematical content is included. You might cite some of the problems students have completed in past years.

You might want to have students give out their questionnaires on the same day. The questionnaires can be given as a homework assignment to save class-time. You should indicate to students what the result of the project will be. For example, will some of the problems be used as part of the year's curriculum. Personally, I would guarantee at least one problem, and perhaps more (depending on the quality of the problems). If the quality is good, then these problems would be very motivating for the students and be an excellent substitution (or addition) to the curriculum. When the projects are complete, have the groups give their oral presentations and collect the written reports.

Evaluation and Processing: Besides whatever comments naturally come out of reviewing all the reports, it seems important in the processing to discuss any surprises that groups encountered in gathering data, particularly if the surprise supports the need to gather data or persist to solve a problem.

Suggested Format: The students work in groups of three and are given a portion of two periods to develop their questions, and a portion of two periods to analyze the data and prepare their oral report. The oral reports and the processing will each require a portion of a class period. The project can be completed in ten to fifteen school days.

Enrichment and Extension. As mentioned, I recommend working on at least one of the problems as a group, or perhaps identify a few problems and divide the class into groups, each addressing one of the selected problems. Also, this is a good problem to assign a second time approximately halfway through the year.

Additional meaningful nonroutine problems

The previous two sections discussed meaningful nonroutine problems that depended directly on input from students identifying problems that were specifically meaningful to them. In this section, I will discuss seven additional nonroutine problems on specific topics that tend to be meaningful for many students and topics not discussed in previous strands.

(1) **"Persistence"**: Using questionnaires, students investigate topics (e.g., How do students define a loving relationship?) that require persistence to research satisfactorily.

Sample student problem statement: Your group task is to learn more about the students in your school. The purpose of this task is to give you practice with a problem requiring persistence. Specifically:

(1) Your group will select three research questions about students in your school to answer. You may select any of the questions from the handout below, PERSISTENCE: RESEARCH QUESTIONS. You may make up one question not on the handout, provided that I judge the question to be as complex and meaningful as the questions on the handout.

(2) Your group will develop a questionnaire that a student can fill out in10 minutes or less that will help you answer your three research questions. I will copy the questionnaire and give it to 30 representative students. For this task, you may assume that the 30 students represent the school.

(3) After the completed questionnaires have been given back, your group will analyze the data from the questionnaire and you will have up to two weeks to complete the task. You will be given time to work in your group up to twice a week. The task will be completed when you believe you have answered the research questions to the best of your ability. You will be permitted to collect additional data; for example, if you wish to develop a second questionnaire, I will give the questionnaire to 30 representative students.

(4) When all the groups are finished, each group will prepare an oral presentation of five minutes or less on the gathered information. I will select the member of your group to present the oral report.

(5) Your group will be required to hand in a written summary of your conclusions including how you organized and compiled your data, and what data supports your conclusions.

Note well: Your choice of topics will not affect the evaluation. Also, the length of time you spend on the project will not affect your grade; that is, if the quality of the work is the same, two groups that spend different amount of time on the task will receive the same number of points.

Teaching suggestions: Go over the directions and answer any student questions. This problem was designed to require persistence to complete well; therefore, it is one of the more difficult problems. You need to be aware whether the students need advice or assistance throughout the task. It might be useful to require a progress report halfway through the project. You might require them to submit the first questionnaire to you before it is administered so that you can give them feedback, if appropriate. It should be clear that the research questions for this task are designed such that even a good questionnaire developed by students would be unlikely to result in the final data needed to draw excellent conclusions; however, the first questionnaire should be well enough constructed to provide enough data to design a second questionnaire to finish the project. When the projects are complete, have the groups give their oral presentations and collect the written reports.

Processing. It seems essential in the processing to focus on the role of persistence in this problem. One way of doing that is to find examples of when data from the first questionnaire provided unclear data that needed to be resolved by constructing a second questionnaire. You might find some examples in the written reports. If not, perhaps asking students for examples during the processing would generate some examples. If the examples are clear enough, you could ask students to think back on previous problems they completed and try to think of examples where persistence was either useful or would have been useful. If not many examples are given, you might assign it as a group brainstorming task or a homework assignment.

Suggested Format. The students work in groups of three. The nature of this problem suggests a different format than usual. My suggestion is that a two-week period be given in which groups meet part of two periods each week on the project. After that, the next problem can be started and students can use some of the group time for that second project to discuss finishing the persistence problem. This may mean allowing a longer portion of a period for group work than usual during the time they are working on two projects. The oral reports and the processing will each require a portion of a class period and should probably occur when all the groups are finish. Note: Two ways that the format can be changed to fit your context are: (a) Have each group research only one or two questions and/or (b) insure that all or most of the questions are researched by at least one group. An alternative is to have the students vote for the questions they are most interested in and eliminate the questions with little or no interest.

Enrichment and Extension: Each group can generate additional questions to research or the class as a whole can select a few questions to investigate as a class.

Persistence: research questions

1 How does having a job influence the life of a student?
2 What are the three improvements in the school that students believe are most important?

3 Are there differences between honor students and other students?
4 Are there differences between athletes and other students?
5 What are the world or political issues most on students' minds?
6 What do students think makes a good teacher?
7 How do students define a loving family?
8 Whom do students consider their heroes/heroines?
9 How do students learn best?
10 What do students do to relax?
11 Where do students feel safe?
12 Why do students have pets?
13 What does it mean to be patriotic?

Comments on the persistence task: I used this task with my Geometry Honors class the last 2 1/2 weeks of school. This was the first "non-geometry" group task that I gave them and, because it was very late in the year, they did not have the time or energy to work on the task as would be ideal. Therefore, their performance was less than what I would have expected (or structured for) under different circumstances. Even with these conditions, I was satisfied with the potential I saw from this field test of this nonroutine problem.

Sequence: The honor students were given the prompt and the directions were reviewed, the only change being that they selected two research questions versus three. They had four class periods to develop a questionnaire which would be administered to a group of random students. During the four periods, they had a task related to the content they were covering to work on when not working on the nonroutine problem. Students used no more than half the four periods to work on the task. This structure seemed appropriate for the task; that is, the groups seem to have enough time to develop a questionnaire of the quality I expected. It should be noted that the task is structured in such a way as to make it likely that students would not develop a questionnaire which would give them clear and quality results; that is, students would have to persist in order to obtain better results. For the second part of the task, students had four additional days to analyze the data, develop and analyze a follow-up questionnaire, and write a report (a little more than a week). The structure of this portion of the task was not appropriate in that students did not have enough time to persist. However, as a result of this experiment, I drew two conclusions: (1) The task is rich in the sense that it requires the student to persist in a relevant setting to obtain quality results. For example, these honor students did not develop initial questionnaires that were likely to generate clear and quality information; therefore they could not complete the task successfully without persisting, the intent of the problem. (2) For the task to be effective I believe that at a minimum the students need a month to work on the second portion of the task. During this time the students could be working on other work and require minimal class-time to meet in groups on the task, ideal might be if they had another group task they were working on that would allow them the flexibility to take some time for the nonroutine problem at their discretion. Also, it might be necessary to process this task and have the students research additional questions in order to get satisfactory results. I say this because, even though students did not have enough time to complete the task, it seemed to me that the students felt that they had completed (or nearly completed) the task satisfactorily. Therefore, it would seem appropriate to process the first reports and allow a second opportunity to demonstrate persistence.

(2) **"Having Fun in Your Community":** Students determine ways to have fun in their local area.

Sample student problem statement: The task for this problem is to discover ways that you and other teenagers your age can have fun in your community or nearby. This task gives you practice solving a problem affecting your life. Specifically:

1   The project will be in two parts. First, your group will identify exactly three ways of having fun in your community or nearby, with one way from each of the following resources: written or internet information, another teenager (not in this class), and an adult. Your group will hand in a short description of the three ideas and give an oral presentation of two minutes or less for this part of the project.

2   Second, after the first part of the project is discussed, your group will research a few specific topics related to having fun in your community or nearby. The topics will be assigned to your group by me with consultation.

3   Each group will prepare an oral presentation of two minutes or less for this part of the project. I will select the member of your group to present the oral report.

4   Your group will turn in a written report for this part of the project, including where you obtained your data.

Teaching suggestions: This problem can be introduced by telling students that this new problem they are going to work on will directly affect their life and, hopefully, illustrate to each student the relevance of problem solving in their life. Go over the directions and try to answer any questions. Be sure that students understand the two stages of completing the project. For the first part of the project, it helps to encourage students to brainstorm resources they might use. Also, make sure that students realize that their task is to come up with the three best ideas, one from each type of resource. Unlike many of the other projects, the groups are limited in how many ideas they can present. When the first part of the project is complete, the class can develop a list of the ideas generated. These ideas can be used by the groups in preparing their reports for the second portion of the project.

For the second portion of the project, divide the topics on the handout HAVING FUN IN YOUR COMMUNITY: TOPICS TO RESEARCH fairly among the groups. Three topics per group is a good number; of course, some topics can be given to more than one group. You may want to change the details of, add or eliminate some topics. For example, instead of researching movies within 3/4 of an hour, it may be more appropriate in your community to investigate within 1 1/2 hours. Also, make it clear that the questions under each topic are not necessarily complete, but rather, suggestive of the questions they should answer for that topic.

One suggested product of this project is a booklet that is made available to all students in the school. If you choose this option, it will probably help to give the students directions in preparing the second part of the report to present their information in a way such that another student reading it would find it useful. When the projects are complete, have the groups give their oral presentations and collect the written reports.

Processing. It seems essential in the processing of this project to focus on the three types of resources (and any other types!) used in this project and what students learned about the value of resources. In addition to sharing relevant sections of group reports, you might ask each student to write a paragraph or two on what they learned about using resources in problem solving. Then the class could discuss the responses. In addition, a discussion of how the problem solving techniques helped improve their lives could be appropriate.

Suggested Format. The students work in groups of three and are given a portion of two periods for the first part and a portion of two periods for the second part. The oral reports and the processing will each require a portion of a class period. The project can be completed in ten to fifteen school days. One option is to make the first part of the project an individual project.

Enrichment and Extension: A good extension is to prepare a summary of the class's findings and circulate the summary among a variety of people (e.g., other students, teachers, community leaders) to solicit additional ideas. Then the findings could be finalized and made available to all students in the community. This extension would also give students a chance to see how persistence can improve the quality of a solution.

Having fun in your community: topics to research

1 Dancing: Where are the three best places in the area for dancing for people under 21 (no drinking)? How can one find out about appearances of live groups or bands with dancing?
2 Movies: What movie theaters are within a 45-minute drive? Where are they publicized? Any special features? Phone numbers?
3 Concerts: Where are there concerts (including smaller places) in the area and how do you find out about them? What is a good way to find out about or purchase tickets?
4 Shopping: What are the best places for shopping within one hour? Identify at least ten stores that would be of interest to teengers.
5 Hiking: Where are the best hiking trails in the area? How do you get information about the trails?
6 Food: Identify at least five restaurants that teenagers might enjoy. Include at least one inexpensive, one moderately priced and one expensive restaurant. What is the best place for pizza, grinders and ice cream nearby and within one hour?
7 Sightseeing: Identify at least ten sightseeing attractions within one hour. Include information concerning cost, directions, and season.
8 Colleges: What colleges are within two hours and how do you find out about events at the colleges? How do you get information about regular events (e.g., football games, etc.)?
9 Amusement Parks: What amusement parks are within two hours? Give information about rides, cost, directions, etc. How do you find out about traveling amusement parks?
10 Professional Sports: Where are the closest places that you can watch professional sport teams compete? How do you get a schedule and tickets? Are there bargains available?

11 Beaches: What are the best beaches locally and within one hour? Include relevant information.

12 Recreational Centers: What kinds of activities do the local Y's, park and recreational centers, and similar organizations offer?

13 Fishing and Hunting: What kind of license do you need to fish and hunt? Where can you get gear? Where are good locations for these activities?

14 Personal Growth: Identify places where teenagers can talk with other teenagers or counselors about personal issues, including alcohol or drug abuse. Are there places or courses for self-improvement?

15 Other Recreation: What other types of recreation are available within one hour? Include activities such as canoeing, waterskiing, golf, skating and skiing. Include relevant information such as cost, directions and a description of what is available.

(3) **"Buying a Car":** Students determine the best new or used car for their values and a given amount of money.

Student problem statement: The task for this problem is to determine the car in a given price range which would be most consistent with what you value in a car. The purpose of this task is to give you practice solving a nonroutine problem involving a real-life application and the use of resources. Specifically:

1 You will select a price range from the choices I give you and determine the car in that price range which would be most consistent with what you value in a car.

2 You will turn in a written report including what resources you used to make your decision, how you arrived at your decision and why you feel your decision is a good one. You should indicate what features and characteristics in a car you value and how your choice is consistent with those values.

3 Only a few students will be asked to orally explain their project, but each student should be prepared to give an oral presentation.

Teaching suggestions: The students need to be able to use a variety of resources to solve the problem. In addition, they will probably need to use specific resources such as Consumer Reports (annual review of used cars and many new cars) and the "blue book" for average prices for cars. If students have had enough success with problems requiring the use of resources, it would be useful to have them attempt this project without introducing them to the specific resources useful for this project. If not, a discussion of a few resources would be useful. I will note that even when you inform them of the resources, this problem usually requires persistence to make the best choice for your values. Go over the directions and answer any student questions. You will need to give them the ranges they can choose for their cars. I suggest a range of at least $3,000 for each choice (e.g., $9,000 to $12,000); that range should allow a variety of cars to be considered. Encourage students to select a range consistent with what they might be able to afford so that their research might be useful. You might want to alert students to the value of using other people as a resource in addition to whatever written resources they use.

You need to decide what prices students can use for the project. One option is to use the prices available in whatever reference material is available locally. You might additionally allow any prices from a newspaper classified advertisement. You should decide whether local businesses that sell automobiles are legitimate resources for the project. If yes, you should coordinate the project with the local businesses. You may need to insure that certain reference materials will be on reserve during the project.

When the projects are complete, have the groups give their oral presentations and collect the written reports.

Processing. Besides whatever comments naturally come out of reviewing all the reports, it seems important in the processing to focus on two topics: (a) Did students use a variety of resources? and (b) Did students persist until reaching a good solution?

Suggested Format. The students work individually on this task; however, you might want to permit two students to work on the problem together if it seems they can work in a way that will result in both students solving the problem effectively. The oral reports, if any, and the processing will require a portion of a class period. The project can be completed in five to ten school days. This problem is probably most appropriate for students at the age that they are already driving or learning to drive.

Enrichment and Extension. Students can research how to buy a car, using resources such as articles from Consumer Reports. The students can put together a booklet for interested high school students, including general hints for looking for and buying a car and listing good buys in different price ranges.

(4) **"Planning and Taking a Class Trip":** Students plan a trip that is inexpensive, interesting and educationally valuable. This problem is like PLANNING A CULTURAL TRIP, without a specific focus on culture and diversity; therefore, the emphasis is more on them

determining the (required) educational focus. Either problem is excellent in helping students see how the process of solving a non-routine problem can result in outcomes better than their initial expectation. I have had many classes plan and take trips that were inexpensive, interesting to them and considered educational by me or the principal. Whenever possible, I integrated this problem into my teaching, always with excellent results!

Student problem statement: The task for this problem is to plan a one-day class trip that will be enjoyable, educational and inexpensive. The purpose of this task is to solve a problem that requires you to use a variety of written and people resources and will increase your appreciation of how problem solving can affect your life in a positive way. Specifically:

1 Each group is to plan a one-day class trip. The trip should have three characteristics: (a) it has educational value as judged by me (or the principal), (b) it is interesting and enjoyable from the point of view of students and (c) it is inexpensive. Your group's plan should include transportation, food considerations, schedule, costs and a description of the day's activities.

2 The first activity in your groups will be to brainstorm what resources your group can use to plan this trip, including written and people resources. In addition, your group can prepare a questionnaire to gather data from the students concerning their interests in various ideas.

3 Each group will prepare an oral presentation of two minutes or less on the gathered information. In your presentation, you are trying to convince the class that your group's proposed trip is the best. I will select the member of your group to present the oral report.

4 Your group will turn in a written report including what resources you used, how you decided on the details of the trip, and a description of the trip.

Teaching suggestions: This project is complex and requires students to find and use a variety of resources. You need to judge whether the students are experienced enough with using resources that they can complete the project successfully. Hopefully, by this point in the curriculum, most students will be prepared for the task. One option is to include a fundraiser from ENVIRONMENTALLY SOUND FUNDRAISING as part of the project; this can be incorporated into the group work or done as a class. Go over the directions and answer any student questions. You might want to emphasize the need to brainstorm and investigate a variety of resources in order to successfully complete the project. The project is best when students know that the class will actually take one of the trips; this helps them directly see the results of their problem solving activities. Depending on your location, you might want to restrict the trip to a major city in your area.

You might want to require at least one questionnaire from each group to gather feedback from students concerning their interest in various ideas. You might want to have the questionnaires all handed out the same time and remind students that ideas on the questionnaires cannot be used by another group. This is a good project to require a written progress report approximately halfway through the project. When the projects are complete, have the groups give their oral presentations and collect the written reports.

Evaluation and Processing. Besides whatever comments naturally come out of reviewing all the reports, it seems important in the processing to discuss how students used resources to solve this problem, including examples of false leads. If the class actually takes the trip, a discussion of how the trip turned out compared to how it was planned could be interesting.

Suggested Format. The students work in groups of three and are given a portion of two periods to start the project and develop their questionnaires, and a portion of two periods to analyze the data and complete the project. The oral reports and the processing will each require a portion of a class period. The project can be completed in ten to fifteen school days. This is a project that it is worth extending the time allotted to insure a good final product.

Enrichment and Extension. After the oral presentations on the trip, the class can plan a trip that combines a number of features from a variety of the proposed trips; this requires good problem solving skills to figure out what the class prefers out of all the ideas. This extension is more feasible if all the trips are to one general location (e.g., a major city). Also, if time permits, a second similar trip would certainly be an excellent extension.

(5) **"Planning a Vacation":** Students plan an enjoyable vacation with a given amount of money and time.

Student problem statement: The task for this problem is to plan a vacation or trip for yourself that would be enjoyable and affordable. The purpose of this task is to solve a problem that requires you to use a variety of written and people resources and will increase your appreciation of how problem solving can affect your life. Specifically:

1   You need to decide what time of the year the vacation or trip will take place; how long the vacation or trip can be; and how much you can afford to spend. In addition, you should add any other conditions that would restrict your choices (e.g., would not want to leave the country; need a vegetarian diet). You should pick conditions that would allow you to actually take this vacation or trip.

2   Each person is to plan the most enjoyable vacation or trip within the given conditions. Your plan should include transportation, food considerations, a tentative schedule, costs and a description of your activities. Your first activity in planning the trip should be to brainstorm what resources you can use to plan this trip, including written and people resources.

3   You will turn in a written report including what resources you used, how you decided on the details of the trip, and a description of the trip.

4   Only a few students will be asked to orally explain their project, but each student should be prepared to give an oral presentation.

Teaching suggestions: This project is complex and requires students to find and use a variety of resources. You need to judge whether the students are experienced enough with using resources that they can complete the project successfully. By this point in the curriculum, students should be able to complete this project with little or no guidance from you. The project can act as part of an evaluation of the effectiveness of the curriculum.

Go over the directions and answer any student questions. You might want to emphasize the need to brainstorm and investigate a variety of resources in order to successfully complete the project. You might want to allow the option of students working in groups on this project, if they intend to do the vacation or trip together. This is a good project to require a written progress report approximately halfway through the project. When the projects are complete, have the groups give their oral presentations and collect the written reports.

Processing. Discuss how students used resources to solve this problem, including examples of false leads. If students take the trip or vacation before the end of school, a discussion of how the trip turned out compared to how it was planned could be interesting.

Suggested Format. The students work individually on this task. The oral reports, if any, and the processing will require a portion of a class period. The project can be completed in ten to fifteen school days. I have assigned this problem effectively with high school seniors and undergraduate students in a problem solving course.

This is an excellent nonroutine problem to follow "Planning and Taking a Class Trip," giving students the opportunity to plan a trip for them (and perhaps a friend) individually, thereby seeing how the approach can help them concretely apply the process in their life now and in the future.

Enrichment and Extension. A second similar trip would certainly be an excellent extension, perhaps planning a weekend (or portion of a weekend) before school is over.

Two additional problems that could be appropriate in some settings:

(1) "Supermarket problem": For this problem the class determines which supermarket in their community (or nearby) they consider to be the best. To start, the class brainstorms features of a supermarket that are important in evaluating its quality. For example, students may come up with categories such as availability of parking, quality of service, price, organization of the supermarket, variety of products and time needed for check-out. After the brainstorming, the class determines about six categories they believe are most important (this can be an interesting math problem!). Then divide the class into groups of three, giving each group enough of the categories so that at least two groups are assigned each of the categories. The task in each group is to develop a procedure (for each of their categories) for evaluating the quality of any supermarket on their categories. For most classes, it helps to require them to be able to assign a value of 1, 2, 3 or 4 (4 the best rating) for each category based on their procedure. Procedures can then be critiqued by either the teacher or other students and revised.

(2) **"Mortgage problem":** In this problem, students first fill out a sheet that identifies their criteria concerning the characteristics of a good mortgage (e.g., projected length of mortgage, location of bank, projected length of ownership, percentage of salary for mortgage). They are then required to investigate the best mortgage available to meet these criteria. To be completed well, this problem requires the student to investigate a variety of written and people resources to discover the mortgage that best fits the established criteria. In addition, a good solution generally requires the student to adjust his/her values to reflect the practical limitations of the options available. Most good solutions require the use of some technology such as a computer spreadsheet or a graphing calculator to investigate the implications of slightly different offers.

If you gave this problem to an average student (without experience solving nonroutine problems) as a level 6 problem (little teacher support), it would be unlikely that the student would be successful and quite possible that the experience would be frustrating. For that student, a level 2 or level 3 structure (much teacher support) would be appropriate to insure that the problem was at the right level of difficulty. In contrast, in the sample curriculum the problem is given as a level 6 problem but is a fourth year problem which means that the students already would have had a variety of experiences solving problems at the easier levels. The point being emphasized is that if we expect students to be able to solve a problem of the complexity of the mortgage problem, we need to provide a curriculum that gradually moves the student from a level at which they find this type of problem frustrating to a level at which this type of problem is almost straightforward.

Assessment

This section of the chapter discusses some of the options for assessing whether the student can demonstrate the ability to apply the steps of solving a nonroutine problem to solving meaningful problems in the student's life. As mentioned already, some problems that are part of the curriculum provide a natural opportunity to assess their ability to apply the steps of solving a nonroutine problem to meaningful problems in the student's life. For example, the four individual problems mentioned in the last section provide a natural opportunity to assess the student's transfer ability. In fact, in practice I have used a student's work on an individual problem to assess their progress and used their work on the problem as an opportunity for instruction. For example, if the student's work on the problem was not complete or satisfactory, I would require them to revisit their solution and resubmit an improved solution, perhaps providing appropriate scaffolding if needed. At times, this requires the student to work beyond the original deadline for the problem. To not overburden the students completing their individual problem, I do not assign additional problems for two weeks or so. Since the individual problems are meaningful to the students, most do not mind spending the extra time on their solution. In addition, processing of the class's work on their individual problems allows for a deepening of their understanding of the three steps, both in terms of the students hearing other student's solutions and you highlighting key examples of the steps.

In my opinion, assessing the students in the context of their work on individually defined meaningful problems is probably the best approach to assess their progress; however, an one period assessment that I have found useful for formative assessment (and a basis for instruction) is a test or assessment that consists of a number of situations that represent potential meaningful problems and asks the students to outline the steps they would (hypothetically) take to solve a few of the problems. The directions emphasize that they are not being asked to identify the "solution" to the problem, but rather to discuss the steps they would take to solve the problem and possible obstacles they might encounter. Of course, a limitation of this approach is the fact that the problems are not actually being solved, just outlined in the limited time of a period. However, I have found this approach very useful in terms of a basis for valuable discussions of the steps in solving meaningful nonroutine problems. Below I have included one version of such an assessment that might provide the basis for you designing your own assessment more relevant to your unique context:

Directions: Select two of the problems listed below to address. For each problem describe the steps you would take to solve the problem. Specifically, identify difficulties or obstacles you would expect; what you would try to solve the problem; and how you would probably need to persist to achieve a good solution. It should be clear that you are not identifying the "solution" or "answer" to the problem (e.g.; the best supermarket is …, or the pet I would get is …); but rather how you would go about solving the problem (e.g., what would your plan be; what people or type of written resources would you use). In documenting your answer for each problem, address the following three areas:

(1) PROBLEM RECOGNITION (30%). Consider the following in your analysis:

What are the probable obstacles in this problem? What is it that makes the problem difficult? Is it clear what the problem is asking for? Is the statement of the problem clear? What might a good solution look like?

(2) TRYING SOMETHING: THE PLAN (50%). Consider the following in writing your plan: Is it appropriate to brainstorm possible approaches to the problem? Are any of the following (or combination of) strategies useful: logical reasoning; guess and check; making a diagram; generating or gathering data; using written resources; talking to other people; trying a new behavior or approach; or letting the problem sit for a while before attempting a solution? What strategies are suggested from step one? What can I do or try to at least get a better idea of what is involved in this problem? Make a list of the steps in your plan.

(3) PERSISTENCE (20%). Describe what you could do in addition to your original plan to insure an excellent solution. Consider the following questions: what might be necessary to complete the solution after you carry out the above plan? How would you know that you have found an excellent solution? What kind of data or information would indicate that I found a good solution?

Problems:

1 You want to take a friend to a special dinner; money is no object. How do you select the restaurant?

2 You see someone rollerblading and decide that you would like to rollerblade. You know almost nothing about rollerblading, how do you get started?

3 You want to cook a special dinner for Mother's Day. You have little experience cooking. How would you go about selecting and preparing the meal?

4 You want to spend a week's vacation in Florida. How would you plan it?

5 You want to buy a good inexpensive running watch. How would you select the best watch?

6 You want a summer job as a waitperson. You want a job with flexible hours and good tips. How do you find the best job?

7 You wish to buy new tires for your car that are a good buy and a good quality. How would you select the best choice?

8 You want to create a garden that would be aesthetically pleasing and do well in a mostly shady area. How would you plan the best garden?

9 You want to identify the best place and time (season) to go whale watching reasonably near to you. How would you identify the best spot?

10 You want to identify the eight best places to camp on a cross-country trip. How would you identify the best campgrounds?

10

Models of implementing a curriculum of nonroutine problems

In this chapter, I will discuss four models of implementing a curriculum of nonroutine problems, each designed for a different context: (a) a mathematics department planning a three or four-year high school mathematics curriculum covering traditional mathematics content, (b) an individual teacher teaching a traditional mathematics course, (c) an individual teacher teaching an elective problem solving course at the middle school, secondary or post-secondary level and (d) a group of educators designing an interdisciplinary curriculum. In this chapter, I will discuss some general considerations that apply to all four contexts. The recommendations in this section are based primarily on my experience field-testing the curriculum with several teachers in a variety of classrooms, including all the traditional high school mathematics courses and a few middle school classrooms (e.g., DeLeon, 2000), as well as an elective one-year course in problem solving I taught from 1984 to 1995 (primarily for seniors who took the course to fulfill their mathematics requirement for graduation) and a semester undergraduate course in problem solving, both of which allowed a more intense focus on nonroutine problems than in a traditional content focused course. The field-testing was supported by a number of summer fellowships, including a ten week Alden B. Dow Creativity Fellowship, during which I wrote a draft of a four-year academic curriculum of 60 nonroutine problems, which became the basis for intense field-testing in one school over a five year period, resulting in the development of a four-year curriculum for all our major mathematics tracks.

DOI: 10.4324/9781003393283-10

Here, I will summarize general guidelines outlined in Chapter 2 that are important when implementing a curriculum of nonroutine problems in each of the four contexts:

(1) The most effective way to improve students' ability to solve nonroutine problems is to repeatedly put them in a situation in which they are given a nonroutine problem at an appropriate level of difficulty to solve (either individually or in cooperative groups), the students work on the problem and generate their best solution, and then you discuss and process the problem as a class.

(2) In an effective sequence it is normal that each problem *is* difficult. It is normal that the student would not be clear about how to solve the problem initially, at times feel as if the problem is not solvable and need to persist until the problem becomes clear.

(3) The teacher needs to provide a structure that allows students to work on problems at an appropriate level of difficulty. For example, we can manipulate the level of support we provide students when solving the problem. In Chapter 2, I define seven levels of support that form an instructional sequence.

(4) A cooperative group model is consistent with the purposes of this curriculum. For example, students benefit from being exposed to the thinking of other students, and cooperative group work provides a supportive atmosphere for dealing with the natural difficulty of the problems.

(5) The curriculum needs to emphasize problems from a variety of fields and problems that are relevant to the student's life. Field-testing indicates that without variety and relevance students are unlikely to transfer the problem solving skills to their day-to-day life. One implication of this guideline is that students with a rich cultural background could benefit from solving problems connected to that background.

(6) My experience indicates that nonroutine problems that do not require significant content prerequisite skills are more effective initial problems than nonroutine problems that require significant content prerequisite skills. This implies that appropriate instruction in solving nonroutine problem does not have to be delayed because of lack of content mastery.

(7) If a teacher feels the need to assign a grade for students' work on nonroutine problems, a method that primarily evaluates the student's quality of effort is reasonably consistent with the pedagogy outlined. In my classes, students that made a reasonable effort

in completing their work received a grade in the 80–89 range (or B range), students that demonstrated exceptional effort or quality of insight received a grade in the 90–100 range (or A range), and students whose efforts were lacking received a lower grade or were required to redo their solution.

In addition, a curriculum of nonroutine problems is not like the traditional mathematics curriculum and cannot be effectively taught with the same expectations. At a certain point in the curriculum, most students feel comfortable with the three steps of a nonroutine problem and even feel a sense of adventure when given the next problem. In contrast, in the beginning of the curriculum, many students feel uncomfortable with the three steps and wish that the problems were "easier" or not part of the curriculum at all. I do not believe that these initial feelings are natural but rather the result of instructional practices over many years that prevent students from experiencing the three steps and, in fact, encourage behaviors contradictory to what is required to solve a nonroutine problem. The curriculum is a gradual process of helping the student feel comfortable with the process of solving a nonroutine problem. For the curriculum to work, it is essential that the teacher understand this aim of the curriculum and feel comfortable supporting the students through the gradual process of changing the way they view mathematics and they solve problems. I can guarantee that the required patience and gentle insistence on the aim of the curriculum will reap a bountiful harvest!

For the first nonroutine problems, it would be wise to spend some time discussing expectations. For example, you might discuss the following four points: (1) The problems require at least a few hours work spread over the period of one to three weeks – they will not be able to answer the problems adequately if they save the assignment for the last day or two. (2) Part of the assignment is to document all nontrivial strategies and solutions attempted as well as the final solution, including justification for selecting the final strategy. (3) There is not one "best" solution for most of these problems, but rather a variety of good solutions they could discover, including some novel solutions. (4) It is quite common with a nonroutine problem that they will "solve" the problem one way and then realize a method to refine the solution. Point out that you encourage this and want them to document it as part of the assignment.

There are probably a variety of ways to effectively structure the assignments. A structure that has worked well for me is to assign one problem approximately every two weeks. Typically, the problem is introduced on a Monday and collected two weeks later. On occasion, I will give the students a week or two off either to allow for a more intense focus on the content of the course or just to give them a break from nonroutine problems. In giving the specific directions for each problem it is important to give the students enough information so that they can engage the problem but, at the same time, not give them so much information that the problem becomes straightforward. If the directions are appropriate, the solution will not be straightforward yet each student will be able to work on the problem without unproductive frustration and will be able to generate a solution. In summary, the problem will be challenging yet not impossible.

The teacher should encourage (or require) the students to work at least some on the problem the first night, emphasizing that this type of start allows ideas to develop even when the student is not directly working on the problem. Also, I like to communicate to students that they are encouraged to use other resources. For example, for each problem I like to make sure that the students realize I am available as a sounding board for ideas. When students do approach me for assistance I certainly do not tell them how to solve the problem but rather try to give the students feedback or hints that will allow them to work productively. Also, they are encouraged to use outside references and resources when appropriate such as the internet, science equipment and math books from previous courses. Of course, a distinction is made between using a resource as an aid in solving a problem and using a resource to solve a problem. It is inappropriate to ask a mathematician for a solution or to try to look up a solution in a book.

Secondary mathematics curriculum

This section of the chapter suggests specific approaches for developing a three or four-year curriculum of nonroutine problems that can be integrated into a typical high school academic mathematics curriculum consistent with the NCTM Standards, including a few sample sequences of problems that constitute a complete curriculum of nonroutine problems.

Before discussing specific sample sequences of problems, I will outline a model for a four-year curriculum of nonroutine problems and guidelines for implementing such a curriculum. I recommend a curriculum consisting of 32 to 60 nonroutine problems (8 to 15 a year) over the three- or four-year high school curriculum. Each problem typically requiring at least one week to complete (one to three hours of class time) and two-thirds of the problems being solved in cooperative groups. Students are required to orally and/or in writing document the process for solving each problem. Approximately 40% of the problems need to involve content not typically considered mathematical, including problems that attempt to increase the student's appreciation of diversity, involve ecology problems with an emphasis on problems directly affecting their lives or are practical applications affecting their lives, including problems that students individually define and solve. I recommend that the curriculum should include an introductory unit that describes the three steps of a nonroutine problem, gives several examples of the steps in a variety of fields and has the students research additional examples. The curriculum of nonroutine problems is designed primarily as one component of a secondary mathematics curriculum (approximately 20% of the allotted time for mathematics instruction) and is easily integrated with a curriculum consistent with the NCTM standards. Specifically, I have found that when implemented well such a curriculum results in as much, if not more, coverage and content achievement as a traditional approach. The curriculum has a positive effect on the student's ability to think like a mathematician and their ability to learn new mathematics.

For example, a representative sequence from a Geometry class (second year, college prep track) included the following eight problems:

(1) "One Inch of Rainfall." Students calculate the volume of water created by one inch of rain in their community.
(2) "Triangles." Students determine under what conditions two triangles are congruent. They are given two, three or four bits of information about the triangle.
(3) "Maximum Area." Students determine in a variety of situations how to maximize area given a certain amount of perimeter.
(4) "Flipping Coins." Students predict the outcome of tossing ten coins 1,000 times (e.g., how many times will the result be 6 heads and 4 tails).
(5) "Graphing 2." Students determine the graphs of some nonlinear equations.
(6) "School and the Environment." Students determine ways to make their school more environmentally sound.
(7) "Planning a Cultural Trip." Students plan a class trip that is inexpensive, enjoyable, and increases the class's appreciation of the diversity of cultures.
(8) "Buying a Car." Students determine the best car for their values and a given amount of money.

In developing a four-year curriculum, attention can be given to gradually shifting the responsibility to students to generate and solve nonroutine problems. For example, it is suggested that the first-year curriculum include an introductory unit that has students study nonroutine problems that others have solved in a variety of fields. In contrast, it is recommended that in the third and fourth year that the curriculum includes a focus on problems generated by the class, as well as a requirement that students individually define and solve significant problems in their own lives. Finally, in the first year of the curriculum, the students need to be introduced to some of the major methods of trying something in real-life problems; for example, using written resources (e.g., Consumer Reports) or internet resources, talking to others or experts and gathering and analyzing relevant data (e.g., develop and administer a questionnaire).

Below, I have included a brief description of two four-year curriculum, the first consisting of 15 problems each year based on my work during the Dow Creativity Fellowship; the second, a curriculum adopted by one high school for its average academic mathematics track, consisting of a minimum of eight problems (except in grade nine which requires six problems and an introductory unit). In developing the second curriculum for one school, we agreed upon requiring a minimum of eight problems a year, with each teacher having a choice of a few problems in a variety of categories. This model satisfied the concern that some teachers had for covering the required curriculum while insuring a reasonable quantity and variety of nonroutine problems to be effective. In addition, at least a few of the problems were directly related to the content of the course. All the problems are described in previous chapters.

One summary of 60 nonroutine problems in a four-year curriculum:

Ninth grade

(1) Getting Acquainted. After completing a questionnaire gathering interesting information about them, students develop a similar questionnaire to gather interesting information about their teacher.

(2) Axioms of Algebra. Students generate a list of algebraic statements that are always true. The project is evaluated on completeness and conciseness.

(3) Scavenger Hunt. Students use a variety of resources to gather certain facts or bits of information (e.g., What size is the largest centipede ever found?). Students must use both written and people resources.

(4) Leaves on a Tree. Students calculate the number of leaves on a large tree. They are required to develop at least three methods.

(5) Interesting Topic. Using written references including articles, books, and interlibrary loans, students research a topic that they are interested in.

(6) Improving the Class. Students generate a list of ways to improve the class based on collected data.

(7) Number of Beans. Students estimate the number of beans in a large jar without touching the jar, using at least three methods.
(8) Calculating the Area of an Irregular Closed Curve. Students calculate (three methods) the area of an irregular closed curve drawn on 1/4″ graph paper.
(9) Comparing Products. Students compare the effect on the environment of certain products (e.g., cloth diapers versus disposable diapers).
(10) Having Fun in Your Community. Students determine ways to have fun in their local area.
(11) Home and the Environment. Students determine how to make their homes more ecologically sound.
(12) Persistence. Using questionnaires, students investigate topics (e.g., How do students define a loving relationship?) that require persistence to research satisfactorily.
(13) Connecting Points. Students devise a procedure for connecting random points (at least 15) on a sheet of paper so as to minimize the distance required to connect them.
(14) Interesting Rituals and Customs. Students find at least five interesting rituals or customs practiced in their community.
(15) Buddha. Students calculate the probability of a sea turtle putting its nose through a floating ring in the Pacific Ocean. The problem is based on a quote by Buddha concerning the probability of being born human.

Tenth grade

(1) Collage. Students create a collage that will interest the students in the class. The problem also helps to build a good atmosphere in the class.
(2) Improving the School. Students prepare a list of five to ten ways to improve the school. They develop questionnaires to help them with the project.
(3) One Inch of Rainfall. Students calculate the volume of water created by one inch of rain in their community.
(4) Triangles. Students determine under what conditions two triangles are congruent. They are given two, three or four bits of information about the triangle.
(5) Planning a Cultural Trip. Students plan a class trip that is inexpensive, enjoyable, and increases the class's appreciation of the diversity of cultures.

(6) Congruency in Quadrilaterals. Students determine under what conditions two quadrilaterals are congruent. They are given four, five or six bits of information about the quadrilateral.

(7) Maximum Area. Students determine in a variety of situations how to maximize area given a certain amount of perimeter.

(8) School and the Environment. Students determine ways to make their school more environmentally sound.

(9) Assigning Grades. Students assign grades given raw scores and two given grades without knowledge of the test or the total number of points on the test.

(10) Predicting the Number of Lunches. Students predict the number of lunches that will be served on a given day. They can only use data collected from students to make the prediction.

(11) Flipping Coins. Students predict the outcome of tossing ten coins 1,000 times (e.g., how many times will the result be 6 heads and 4 tails).

(12) Ecologically Sound Lunch. Students plan three lunches that are ecologically sound, relatively inexpensive, and enjoyable to eat.

(13) M & M's. Students calculate how many M & M's it would take to fill their classroom. They are required to develop at least three methods.

(14) Buying a Car. Students determine the best new car for their values and a given amount of money.

(15) Geometric Constructions. Using student developed measuring devices, the students are given a variety of outdoor geometric constructions to complete. Students do not know what constructions they will be given so they must prepare for a variety of problems.

Eleventh grade

(1) Awesome Tape. Students develop a half hour tape of music that the class will like.

(2) Dividing a Line Segment. Students devise a method to divide a line in a manner that embodies the Golden Mean.

(3) Environmentally Sound Fundraising. Students determine a variety of ways to raise money for school groups that are environmentally sound.

(4) Planning and Taking a Class Trip. Students plan a trip that is inexpensive, interesting, and educationally valuable.

(5) Problem 1. Each individual student generates a problem of interest and investigates it.

(6) Expensive Tape. Students determine three box designs that minimize the need for an expensive tape to wrap it.
(7) R(x). Students are given the abstract definition of a function (R(X) = R(X −1) + X) and one value of the function and asked to determine a variety of difficult values (e.g., R(1000)).
(8) Area Under a Curve. Students determine the area bounded by

$$y = x^2 + 2, y = 0, x = 0 \quad \text{and} \quad x = 3.$$

(9) Favorite Television Shows. Students predict the favorite television shows of students through the use of a questionnaire.
(10) Ecologically Sound Community. Students determine a variety of ways to make their community more environmentally sound.
(11) Problem 2. Each individual student generates a problem of interest and investigates it.
(12) Finding a speaker. Students find an inexpensive, interesting speaker on the topic of improving appreciation of cultural diversity.
(13) F(x). Students are given the abstract definition of a function (F(ab) = F(a)F(b)) and three values of the function and asked to determine a variety of difficult values.
(14) Making Predictions. Students predict the outcome of certain experiments involving picking objects out of a bag containing 20 green objects, 10 red objects and 5 blue objects (e.g., probability of picking two green objects in a row).
(15) Slope. Students develop a method to determine the slope of a variety of nonlinear functions.

Twelfth grade
(1) Generating Nonroutine Problems. Students generate a list of nonroutine problems that the class would be interested in studying.
(2) G(x). Given that G(ab) = G(a) + G(b) and three values of G(x), students determine the values of G(x) for x = 1 to 20. Averaging is an unacceptable method.
(3) Marking G(x). Students determine a reliable, yet time saving method for grading the G(x) problem.
(4) Improving the Community. Students determine actions which would improve the quality of the community in which they live.
(5) Increasing Diversity. Students determine how an environmental group can increase the diversity of its membership and better serve a more diverse group of people.

(6) Making an Area Equal to One. Students calculate the value of b such that the area bounded by y = 1/x, y = 0, x = 1, and x = b is equal to 1, 2 and 1,000.

(7) Maximum Volume. Given a piece of construction paper, the students construct the solid without a lid that holds the most volume.

(8) Profit and the Environment. Students determine a method for calculating profit that takes into account the effect on the environment.

(9) Problem 3. Each individual student generates a problem of interest and investigates it.

(10) The Best Menu for the School. Students predict the most popular menu for the school based on a questionnaire they develop and administer to a portion of the student body.

(11) Environmentally Sound Products. Students develop a list of ten products that they believe would have the most positive effect on the environment if made available to people in the community.

(12) Approximating Pi. Students calculate pi in ten different ways. Students are evaluated on accuracy, creativity and variety.

(13) Problem 4. Each individual student generates a problem of interest and investigates it.

(14) Exploring Functions. Students compare a variety of functions (e.g., log x, x squared, 2 to the x power) in a written project.

(15) Planning a Vacation. Students plan an enjoyable vacation with a given amount of money and time.

The second curriculum developed for one high school which requires only 8 problems per year (versus 15 above) and consists of the following choices:

Ninth grade (algebra 1)

The curriculum for ninth grade should minimally consist of the introductory unit on nonroutine problems and six nonroutine problems consisting of:

(1) The following two problems:

Buddha. Students calculate the probability of a sea turtle putting its nose through a floating ring in the Pacific Ocean. The problem is based on a quote by Buddha concerning the probability of being born human.

Graphing 1. Students determine the graphs of some nonlinear equations by substituting easy values and persisting until the graphs are clear.

(2) Two of the following problems:

Leaves on a Tree. Students calculate the number of leaves on a large tree. They are required to develop at least three methods.

Number of Beans. Students estimate the number of beans in a large jar without touching the jar, using at least three methods.

Calculating the Area of an Irregular Closed Curve. Students calculate (three methods) the area of an irregular closed curve drawn on 1/4" graph paper.

Five Calculations. Students devise a method for reducing random numbers between 100 and 900 in five operations (multiply, divide, add and/or subtract) using only the whole numbers 1 to 9. The problem requires them to work effectively with limited time and resources.

(3) Two of the following problems:

Improving the Class. Students generate a list of ways to improve the class based on collected data.

Having Fun in Your Community. Students determine ways to have fun in their local area.

Home and the Environment. Students determine how to make their homes more ecologically sound.

Persistence. Using questionnaires, students investigate topics (e.g., How do students define a loving relationship?) that require persistence to research satisfactorily.

Interesting Rituals and Customs. Students find at least five interesting rituals or customs practiced in their community.

Comparing Products. Students compare the effect on the environment of certain products (e.g., cloth diapers versus disposable diapers).

(4) In addition to the six problems, if there is time the teacher may include some or all of the following:

Getting Acquainted. After completing a questionnaire gathering interesting information about them, students develop a similar questionnaire to gather interesting information about their teacher.

Scavenger Hunt. Students use a variety of resources to gather certain facts or bits of information (e.g., What size is the largest centipede ever found?). Students must use both written and people resources.

Interesting Topic. Using written references including articles, books, and interlibrary loans, students research a topic that they are interested in.

Tenth grade (geometry)

The curriculum for tenth grade should minimally consist of eight nonroutine problems consisting of:

(1) The following five problems:

One Inch of Rainfall. Students calculate the volume of water created by one inch of rain in their community.

Triangles. Students determine under what conditions two triangles are congruent. They are given two, three or four bits of information about the triangle.

Maximum Area. Students determine in a variety of situations how to maximize area given a certain amount of perimeter.

Flipping Coins. Students predict the outcome of tossing ten coins 1,000 times (e.g., how many times will the result be 6 heads and 4 tails).

Graphing 2. Students determine the graphs of some nonlinear equations by substituting easy values and persisting until the graphs are clear.

(2) One of the following problems:

School and the Environment. Students determine ways to make their school more environmentally sound.

Ecologically Sound Lunch. Students plan three lunches that are ecologically sound, relatively inexpensive, and enjoyable to eat.

(3) One of the following problems:

Interesting Rituals and Customs. Students find at least five interesting rituals or customs practiced in their community. (If not completed in ninth grade.)

Planning a Cultural Trip. Students plan a class trip that is inexpensive, enjoyable, and increases the class's appreciation of the diversity of cultures.

(4) One of the following problems:

Improving the School. Students prepare a list of five to ten ways to improve the school. They develop questionnaires to help them with the project.

Predicting the Number of Lunches. Students predict the number of lunches that will be served on a given day. They can only use data collected from students to make the prediction.

Buying a Car. Students determine the best new car for their values and a given amount of money.

(5) In addition to the eight problems, if there is time the teacher may include some or all of the following:

Collage. Students create a collage that will interest the students in the class. The problem also helps to build a good atmosphere in the class.

Assigning Grades. Students assign grades given raw scores and two given grades without knowledge of the test or the total number of points on the test.

Geometric Constructions. Using student developed measuring devices, the students are given a variety of outdoor geometric constructions to complete. Students do not know what constructions they will be given so they must prepare for a variety of problems.

Honors:

Congruency in Quadrilaterals. Students determine under what conditions two quadrilaterals are congruent. They are given four, five or six bits of information about the quadrilateral.

Eleventh grade (algebra 2)

The curriculum for 11th grade should minimally consist of eight nonroutine problems consisting of:

(1) The following four problems:

Expensive Tape. Students determine three box designs that minimize the need for an expensive tape to wrap it.

R(x). Students are given the abstract definition of a function $(R(X) = R(X-1) + X)$ and one value of the function and asked to determine a variety of difficult values (e.g., $R(1000)$).

F(x). Students are given the abstract definition of a function $(F(ab) = F(a)F(b))$ and three values of the function and asked to determine a variety of difficult values.

Making Predictions. Students predict the outcome of certain experiments involving picking objects out of a bag containing 20 green objects, 10 red objects and 5 blue objects (e.g., probability of picking two green objects in a row).

(2) One of the following problems:

Ecologically Sound Community. Students determine a variety of ways to make their community more environmentally sound.

Environmentally Sound Fundraising. Students determine a variety of ways to raise money for school groups that are environmentally sound.

(3) Two of the following problems:

Individual Problem 1. Students generate a problem of interest and investigate it.

Individual Problem 2. Students generate a problem of interest and investigate it.

Planning and Taking a Class Trip. Students plan a trip that is inexpensive, interesting, and educationally valuable.

(4) One of the following problems:

Finding a speaker. Students find an inexpensive, interesting speaker on the topic of improving appreciation of cultural diversity.

Planning a Cultural Trip. Students plan a class trip that is inexpensive, enjoyable and increases the class's appreciation of the diversity of cultures. (May have done in tenth grade.)

(5) In addition to the eight problems, if there is time the teacher may include some or all of the following:

Dividing a Line Segment. Students devise a method to divide a line in a manner that embodies the Golden Mean.

Area Under a Curve. Students determine the area bounded by $y = x^2 + 2, y = 0, x = 0$ and $x = 3$.

Favorite Television Shows. Students predict the favorite television shows of students through the use of a questionnaire.

Slope. Students develop a method to determine the slope of a variety of nonlinear functions.

Twelfth grade (advanced math/precalculus)

The curriculum for twelfth grade should minimally consist of eight nonroutine problems consisting of:

(1) The following three problems:

G(x). Given that $G(ab) = G(a) + G(b)$ and three values of $G(x)$, students determine the values of $G(x)$ for $x = 1$ to 20. Averaging is an unacceptable method.

Maximum Volume. Given a piece of construction paper, the students construct the solid without a lid that holds the most volume.

Exploring Functions. Students compare a variety of functions (e.g., log x, x squared, 2 to the x power) in a written project.

(2) Three of the following problems:

Generating Nonroutine Problems. Students generate a list of nonroutine problems that the class would be interested in studying.

Improving the Community. Students determine actions which would improve the quality of the community in which they live.

Individual Problem 3. Students generate a problem of interest and investigate it.

Individual Problem 4. Students generate a problem of interest and investigate it.

Planning a Vacation. Students plan an enjoyable vacation with a given amount of money and time.

(3) One of the following problems:

Increasing Diversity. Students determine how an environmental group can increase the diversity of its membership and better serve a more diverse group of people.

Finding a speaker. Students find an inexpensive, interesting speaker on the topic of improving appreciation of cultural diversity.

Planning a Cultural Trip. Students plan a class trip that is inexpensive, enjoyable and increases the class's appreciation of the diversity of cultures.

(4) One of the following problems:

Profit and the Environment. Students determine a method for calculating profit that takes into account the effect on the environment.

Environmentally Sound Products. Students develop a list of ten products that they believe would have the most positive effect on the environment if made available to people in the community.

(5) In addition to the eight problems, if there is time the teacher may include some or all of the following:

Marking G(x). Students determine a reliable, yet time saving method for grading the G(x) problem.

Making an Area Equal to One. Students calculate the value of b such that the area bounded by $y = 1/x$, $y = 0$, $x = 1$, and $x = b$ is equal to 1, 2 and 1,000.

The Best Menu for the School. Students predict the most popular menu for the school based on a questionnaire they develop and administer to a portion of the student body.

Approximating Pi. Students calculate pi in ten different ways. Students are evaluated on accuracy, creativity and variety.

Connecting Points. Students devise a procedure for connecting random points (at least 15) on a sheet of paper to minimize the distance required to connect them.

One-year curriculum for teaching a traditional mathematics course

The model for a four-year curriculum needs to be adjusted slightly when teaching a one-year traditional academic course (e.g., Algebra 1) at the middle school level, high school level or post-secondary level, and it is unlikely that the students will have a curriculum of nonroutine problems the following year. In this type of situation, based on my experience and the experience of teachers with whom I have worked, it is difficult to reach the point that students will be able to demonstrate the ability to solve nonroutine problems in their lives. For example, I found that with my work with Calculus students, generally they needed at least a sequence of eight problems before they could demonstrate the ability to use the three steps effectively to solve nonroutine mathematical problems but they may not necessarily be able to transfer the skills to "non-mathematical" problems in their life. On the other hand, a seventh grade teacher (DeLeon, 2000) that field-tested a curriculum of ten nonroutine problems for a master's project gathered data that demonstrated that the students were better able to discuss possible solutions to real-life nonroutine problems than a control group. He also demonstrated how the student's approach to problem solving had significantly changed by the end of the experiment. In summary, my experience indicates that even within the context of a one-year traditional academic course, the students' ability to solve nonroutine problems and do mathematics can be significantly affected; however, you cannot expect that the average student would be able to define and solve significant nonroutine problems in their life. In addition to the guidelines already mentioned, I suggest that in such a context you (a) try to include at least 12 nonroutine problems in your curriculum; (b) start with introductory problems (as previously described; e.g., the ten problems in Chapter 3) that, with support, the students can solve successfully and begin to concretely see the workings of the three steps; (c) have the students complete the introductory unit described elsewhere and (d) include assignments in which you give the students a choice of a few real-life situations involving nonroutine problems and have them describe how they would solve them (e.g., planning an inexpensive and enjoyable vacation, obtaining a good mortgage for certain given conditions). One problem I particularly like in the context of a one-year curriculum is the problem of planning and taking a one-day trip that is inexpensive, educationally valuable (as judged by a panel of teachers or the principal) and enjoyable. When appropriately structured, the students are always surprised that they can plan a trip that is much less expensive than the traditional school trip *and* more enjoyable. They concretely see how they can apply the three steps to solve a meaningful problem in a way far beyond their expectations. The sample sequences of problems that are described above provide an appropriate starting point for selecting problems for this one-year model.

For a middle school class I would suggest starting by selecting problems that seem appropriate from two groups of problems, (1) the ten introductory problems from Chapter 3 and (2) the suggested problems from the lists for ninth and tenth grade. I would suggest that your problems include at least a couple of problems from the non-mathematical strands, helping to facilitate transfer to meaningful real-life problems, for example the non-mathematical problems from the ninth and tenth grade suggestions. Also, as implied previously, planning a one-day trip that the class actually takes that is enjoyable, inexpensive and educational is certainly an excellent problem to include.

For an undergraduate mathematics course for liberal studies, your choice of problems depends on the context and goals of the course. For the course I taught the major goal was to be able to solve nonroutine problems; therefore, I was free to select the problems that I felt were most appropriate for that goal, versus concern over covering traditional content objectives. In that situation I selected mathematical problems that required few, if any, content prerequisite skills, such as the problems in Chapter 3, to insure a strong focus on solving nonroutine problems versus a focus on mathematics content. Additional mathematics problems can be selected from the ninth to twelfth grade problem lists, depending on which problems you believe the students will have the necessary prerequisite skills or problems for which you will be willing (and have the time) to teach the prerequisite skills as part of the course. In addition, the course should integrate a variety of nonroutine problems involving non-mathematical content, perhaps selected from the ninth to twelfth grade problem lists. In addition, I assigned and would recommend that they identify and solve at least two nonroutine problems in their life that are meaningful, assigned after they have experience solving at least a few nonroutine problems. As mentioned previously, when working with the students on the individual problems I allow flexibility in how I work with each student, focusing on what is needed to best support the student in making the transfer to their lives. For example, a few students might need additional support to understand the steps in solving a real-life meaningful nonroutine problem; that is, their initial solution may not be consistent with the quality you expect to demonstrate mastery. In that type of situation, I try to provide appropriate scaffolding and give them additional time to redo the problem. Depending on the situation, I might only require one individual nonroutine problem rather than two for those students, hoping that the additional work on the one problem will facilitate mastery for transfer to meaningful personal nonroutine problems. I might also allow for work on a third individual problem for extra credit or require it for earning a grade of A.

An individual teaching an elective problem solving course

The two models discussed already in this chapter apply to contexts relevant to most teachers; that is, contexts in which the teaching of how to solve nonroutine problems is expected to be integrated into a traditional content oriented mathematics curriculum. However, in my experience, that type of context de-emphasizes the significance of instruction in the process of problem solving and makes it difficult to maintain a focus on the process. In contrast, I have found that in a one-year course in which the focus is on the process of problem solving, most students demonstrate the ability to solve nonroutine problems involving traditional mathematical contexts and the ability to define and solve nonroutine problems in their own life. For example, in the one-year secondary elective course that I taught, by the second half of the course, students were required to identify and solve at least two nonroutine problems in their life. In addition, the few students that were not seniors and took additional mathematics courses in high school reported improved ability to understand the mathematics in their next course compared to previous years. Therefore; my recommendation for a high school curriculum is that the first year of instruction be a course with the major focus on solving nonroutine problems (of course, other approaches that focus on the process of problem solving, perhaps from a different framework, could be appropriate). I would include problems from the traditional content areas that did not require significant content prerequisite skills (see the examples mentioned previously) as well as at least 40% of the problems from strands not typically associated with the mathematics curriculum. My hypothesis is that such a course would not only teach the process of solving nonroutine problems but also would prepare students to "cover" the content typically covered in four years of high school in three years; that is, they learn how to think as a mathematician. Philosophically, I need to make two comments. First, a K-8 mathematics curriculum with an appropriate emphasis on the process of problem solving might make such a course unnecessary, but unfortunately, such a K-8 curriculum is very rare. Second, I do not believe that there is a good rationale for much of the content that we now include in the mathematics curriculum; that is, if I was designing what I believed would be an effective secondary mathematics curriculum, I would not include a portion of the content objectives that are presently common in the curriculum.

I realize that such a course for first year public high school students would not be a realistic possibility in most (if not all) public school systems. Indeed, I did not attempt to implement such a course in the mathematics department that I chaired! However, I was able to implement such a one-year elective course. I will note that the course originally was meant for average academic students to take in addition to their traditional academic mathematics course to strengthen their ability to learn new mathematics in future academic mathematics courses. However, with few exceptions, it became a course mainly for senior students who needed one more mathematics course to graduate and were unlikely to pass another traditional academic course, due to lack of motivation and/or ability.

For such a one-year elective course I would recommend a similar approach to the recommendations for the above section "One-year curriculum for teaching a traditional mathematics course," especially the recommendations for an undergraduate course, including requiring few content prerequisite skills; at least 40% of the problems from what might be labelled as "nonmathematical" content; and requiring at least two individual meaningful personal problems.

A group of educators designing an interdisciplinary curriculum

Finally, I have completed some work in defining alternative models for secondary education (e.g., 1996). In that work, one component of the model focuses on the process of problem solving in an integrative manner. Specifically, students take a required course in the process of problem solving which includes a variety of content strands (e.g., the elective course described above, perhaps with less problems in traditional mathematics strands) and then apply what they learn in a variety of community contexts (that they generate or select from suggested sites). For example, students can work individually or in small groups with local businesses, nonprofit organizations, local artists or writers, or local government (e.g., a conservation committee to review open space needs of the community). Individual work in community contexts would be the basis of discussion in the required core course. In addition, the class could define/identify nonroutine problems that the whole class could address. A curriculum grid can be developed to insure that students work in a variety of contexts over the course of their education. It should be mentioned that another component of the model suggests a process for identifying core skills required of all students and monitoring their progress in mastering those skills.

Conclusion

I have tried in this book to provide clearly all the information and guidelines you need to implement a sequence of nonroutine problems; however, I know that actual implementation in your context can be more complicated than the models and guidance given here. Therefore, I encourage you to contact me if you have a question concerning implementation at rlondon@csusb.edu.

References

DeLeon, A. (2003). *A curriculum of nonroutine problems in the middle school.* Draft of master's project. California State University, San Bernardino.

London, R. (1996, March). *An alternative school model.* Paper presented at the Association for Supervision and Curriculum Development National Conference, New Orleans.

11

Additional nonroutine problems and resources

In this last chapter, I discuss additional material not considered appropriate for earlier chapters but hopefully useful to at least some readers. Specifically, I will discuss three topics, additional nonroutine problems I have field-tested that would not be appropriate for most readers' contexts, additional problems that I have not implemented but I believe they might be useful for some readers and, finally, some other resources that may be useful in implementing a curriculum of nonroutine problems, including an outline of an introductory unit.

Additional field-tested nonroutine problems

In this section, I discuss five additional problems that I have only field-tested with an honors Geometry class or my AP Calculus class and I do not believe that they would be appropriate for most teachers. However, if there are a few teachers interested I have included a summary of my experience with them and my recommendations for implementation.

   Problem 1. "Congruency in quadrilaterals": This is a more advanced nonroutine problem that I have only field-tested with honors ninth grade geometry students, on track for calculus as seniors. Typically, these students already completed the congruency in triangles nonroutine problem, as well as additional nonroutine problems. Even with their excellent mathematics ability and experience solving nonroutine problems, this is a very challenging problem, that unlike the concept of congruency in triangles is not part of any

DOI: 10.4324/9781003393283-11

high school geometry textbook or curriculum I have seen. Due to the difficulty of the problem, when I have field-tested this problem, I have integrated an introductory worksheet that provides a foundation for establishing a reasonable basis to engage with the problem at an appropriate level of difficulty. I will note that despite the difficulty of the problem I have found it to be a valuable component of a curriculum of nonroutine problems for these students!

Since it is unlikely that the reader will have solved this problem, one solution for this nonroutine problem is offered below that should make the directions and teaching suggestions more understandable:

(1) *Five Bits of Information*: In theory, there are 30 different methods to combine the sides and angles of a quadrilateral into five bits of information (e.g., SASAS, AASAS, etc.). Although some of these methods describe essentially equivalent situations (e.g., SSSAS and ASSSS), it is difficult to investigate the 30 methods as one group. We will simplify the process of investigating these methods by dividing them into four groups based on the number of sides given:
   a) *Four sides, One Angle*. No method involving four sides and one angle guarantees congruency.
   b) *Three Sides, Two Angles*. There are ten different methods of combining three sides and two angles, six of which represent different situations. Only one of these methods, SASAS, guarantees congruency.
   c) *Two Sides, Three Angles*. There are ten different methods of combining two sides and three angles. However, if one realizes that when you are given three angles of a quadrilateral, the fourth angle can be easily determined, then there are only two situations which need to be explored: (1) The two given sides are adjacent (SSAAA, SASAA, SAAAS, ASSAA, ASASA, AASSA, AASAS and AAASS). These methods are valid. (2) The given sides are opposite each other (SAASA and ASAAS). This method does not guarantee congruency.
   d) *One Side, Four Angles*. It should be noted that this category is the same as one side and three angles (fourth angle can be determined from the other three). This method does not guarantee congruency.
(2) *Four and Six Bits of Information*: There is no valid method for establishing congruency between two quadrilaterals with only four bits of information. This can be established by realizing that neither of the two methods which work with five bits of information work if one bit of information is eliminated.

Are there any methods using six bits of information which cannot be reduced to a valid method using five bits of information? For example, SASASA would not qualify because this method is equivalent to SASAS (a valid method) with an angle added. This question is easily answered by considering the number of sides: (1) two sides, four angles – reducible to two sides, three angles, (2) three sides, three angles – reducible to two adjacent sides, three angles (valid method) and (3) four sides, two angles – four sides, two adjacent angles is reducible to SASAS; however, *four sides, two opposite angles is a valid method that is not reducible to any valid method using five bits of information.*

Finally, a valid approach using five bits of information is one angle formed by intersecting diagonals of a quadrilateral and the length of the four portions of the two diagonals.

Sample directions for the students for this nonroutine problem are as follows:

(1) Recall that two polygons are congruent only if all corresponding angles and sides of the two polygons are congruent. For example, two triangles are congruent only if all six pairs of corresponding parts (three angles, three sides) are congruent. Further, there are a variety of techniques for proving two triangles congruent by demonstrating that only three pairs of corresponding parts are congruent. These techniques include: SSS (three sides of one triangle are congruent to three sides of another triangle), SAS (two sides and the included angle of one triangle are congruent to the corresponding two sides and the included angle of another triangle), ASA (two angles and the included side…) and AAS (two angles and a not-included side…). Also, two techniques do not insure congruency: SSA (two sides and a not-included angle…) and AAA (three angles…). For this exercise, you will explore under what conditions two quadrilaterals are congruent. For each situation you are given you will conclude either YES, two quadrilaterals satisfying these conditions must be congruent or NO, two quadrilaterals satisfying these conditions may not be congruent. You need to support your conclusion with a diagram, written explanation, or a combination of a diagram and written explanation.

In addition, you need to recall the following information: (1) the sum of the measures of the interior angles of a quadrilateral is 360 degrees, and (2) a quadrilateral is a polygon with exactly four sides, including quadrilaterals which are not convex.

(2) Investigate the following: given a quadrilateral with five parts (consisting of angles and sides) congruent to the corresponding five parts of a second quadrilateral, under what conditions must the two

quadrilaterals be congruent? Under what conditions are the two quadrilaterals not necessarily congruent? Support each of your conclusions with either a diagram, written explanation, or both. Your work will be evaluated for each of the following categories: (a) comprehensiveness – did you consider all the possible cases? (b) correctness – are your conclusions correct? (c) support for your conclusions – is the support for conclusions valid and adequate? and (d) organization – did you find an elegant or concise method for organizing your data and conclusions? It should be noted that you are not restricted to organizing your data or conclusions as in previous work (e.g., congruency for triangles).

(3) Also, investigate the following: (a) given a quadrilateral with four parts (consisting of angles and sides) congruent to the corresponding four parts of a second quadrilateral, are there any conditions under which the two quadrilaterals must be congruent? and (b) given a quadrilateral with six parts (consisting of angles and sides) congruent to the corresponding six parts of a second quadrilateral, are there any conditions under which the two quadrilaterals must be congruent *and* the conditions cannot be reduced to a valid method using five parts (e.g., ASASAS is valid but can be reduced to ASASA, a valid method with five bits of information).

Teaching suggestions: This problem requires a one to two week period with the students having at least three partial periods to work in groups. If time permits, oral presentations on the projects would be desirable. Before they start the investigations above, I illustrate quadrilaterals that are not convex. In addition, I integrate one to three introductory worksheets to help insure that the nonroutine problem on congruency in quadrilaterals is at the right level of difficulty. Without the introductory activities, the nonroutine problem could be too difficult (unnecessarily frustrating) to appropriately engage most geometry students. Each teacher needs to decide how many worksheets, if any, are appropriate for his/her class. My experience is that one worksheet is adequate for an honors geometry class with some experience with group problem solving. One introductory worksheet is given below, followed by the solutions for the worksheet:

CONGRUENT QUADRILATERALS
WORKSHEET
GROUP NUMBER:_____NAMES:_____

Directions: For each of the three problems given, determine whether two quadrilaterals satisfying the given conditions must be congruent. Support your conclusion with either a diagram, written explanation, or both.

Note: For this worksheet "AB = 5 inches" means the length of the line segment AB is five inches (versus the line through A and B), "quadrilateral ABCD is congruent to quadrilateral EFGH" i.e. m<A = m<E, etc.

(1) Four angles and one side of a quadrilateral are congruent to the corresponding four angles and one side of another quadrilateral.

(2) Four sides and one angle of a quadrilateral are congruent to the corresponding four sides and one angle of another quadrilateral. For example, AB = EF; BC = FG; CD = GH; DA = HE; and m<(1) = m<(5).

(3) ASASA. For example, m<(1) = m<(5); AB = EF; m<(2) = m<(6); BC = FG; and m<(3) = m<(7).

The solutions for this worksheet are as follows:

(1) Four angles and one side of a quadrilateral are congruent to the corresponding four angles and one side of another quadrilateral. For example, m<(1) = m<(5); m<(2) = m<(6); m<(3) = m<(7); m<(4) = m<(8); and AB = EF.
NO; e.g., two rectangles with two sides equal (opposite sides) but not the second set of opposite sides.

(2) Four sides and one angle of a quadrilateral are congruent to the corresponding four sides and one angle of another quadrilateral. For example, AB = EF; BC = FG; CD = GH; DA = HE; and m<(1) = m<(5).
NO; the "second" quadrilateral is not convex.

(3) ASASA. For example, m<(1) = m<(5); AB = EF; m<(2) = m<(6); BC = FG; and m<(3) = m<(7).
YES; notice the fourth set of angles can be determined.

Problem 2. "Making an Area Equal to One": Students calculate the value of b such that the area bounded by $y = 1/x$, $y = 0$, $x = 1$ and $x = b$ is equal to 1, 2 and 1,000.

This problem is typically later in the curriculum (for students with experience solving nonroutine problems; e.g., my calculus class after solving at least a few nonroutine problems) and explores the area bounded by $y = 1/x$, $y = 0$, $x = 1$ and $x = b$ and requires the student to answer three questions: What values of b will yield an area of 1 (see Figure 11.1), 2 and 1,000 (answers: e, e^2 and e^{1000})?

Sample student problem statement: The task for this problem is to calculate the value of t such that the area of the region illustrated in (Figure 11.1, without $x = e$) will be 1, 2 and 1,000. The purpose of this task is to give you practice solving a nonroutine problem involving area and being careful. Specifically:

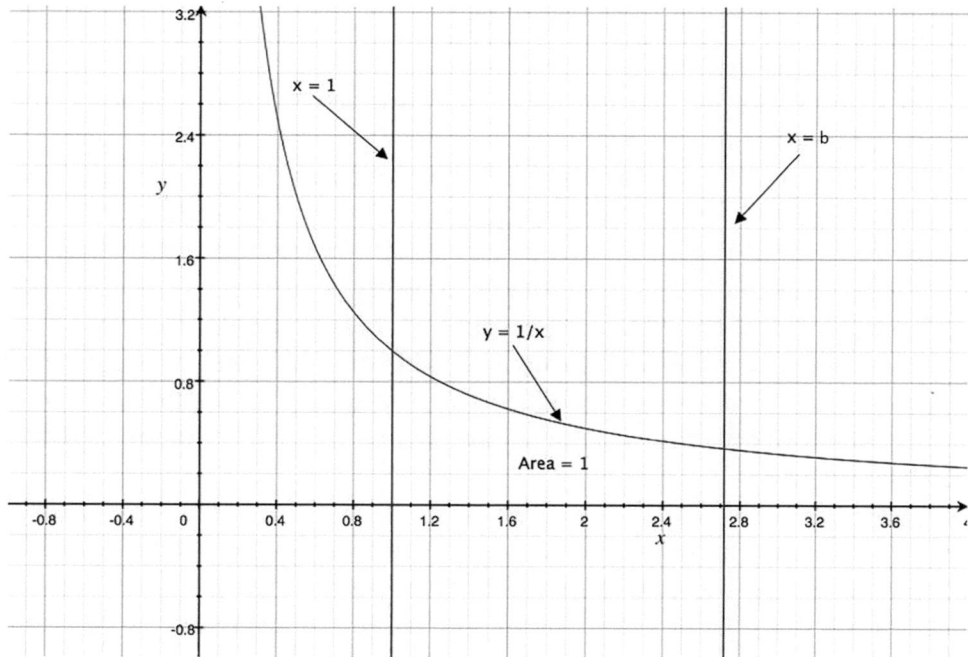

Figure 11.1 Area bounded by y = 1/x, y = 0, x = 1 and x = b.

(1) Figure 11.1 is a sketch of a region bounded by the graphs: y = 1/x, y = 0, x = 1 and x = t. Find the value of t such that the area of the region is one, two, and 1,000. If necessary, round off your answer to the nearest 1000th. For an area of 1,000, you may write your answer as a non-simplified arithmetic expression; e.g., 9876 × 4873/45 or 1234^{45}.

(2) You will turn in a written report including how you made your calculation and why you chose that method. You should describe all methods you considered.

(3) Only a few students will be asked to orally explain their project but each student should be prepared to give an oral presentation.

Teaching suggestions: One way to help students get started and also to be sure they understand the problem is to present the following rough estimate of t for an area of one square unit: So as not to unnecessarily confuse the students, first review the formula for the area of a trapezoid and be sure they realize that A and B are both trapezoids. Then you can calculate the area of A = 1/2(1 + 1/2)1 = 3/4 and the area of B = 1/2(1/2 + 1/3)1 = 5/12, where A and B are the trapezoids circumscribing the graph going through the points (1,1), (2,1/2) and (3,1/3). Through discussion, students should realize that the value is between 2 and 3. Indicate to students that this is a very rough estimate of the value and that their assignment is to determine the value of t for each of the three areas as best as they can.

Let students know that the values of t are not rational; therefore, they are attempting to find the best rational approximation for t. You might add that it is unlikely (but not impossible!) that they will calculate t to the nearest 1000th and that their task is to do the best they can in the given time. It may be helpful to give an analogy of a job situation in which time is a factor; e.g., teaching in the sense that you never have time to plan the "perfect" lesson or unit and you need to do your best given the time and other restraints. A mathematical example is that in real life a perfect circle cannot be constructed; therefore, in industries such as aerospace the task is to build the closest approximation to a circle (e.g., as in the construction of a cylinder such as a piston).

Students should realize that they cannot figure out the value of t for an area of 1,000 square units directly; i.e., the number is much too large (e^{1000}). It is not necessary, but you might want to imply that the problem is to discover a pattern. It is important *not* to suggest that more than two examples are needed to be confident of a pattern, an important focus of the discussion section is whether students generated more than two examples before concluding that there was a pattern. It is suggested for this problem that the students not be allowed to use computers or programmable calculators. The rationale for this restriction is that a major focus of the problem is the manipulative skills, especially the ability to organize and work with complicated calculations. One of the extension exercises reinforces these skills. If necessary, remind students that they need to include their calculations in the documentation. It would be helpful to provide graph paper for their use. When the projects are complete, have the groups give their oral presentations and collect the written reports.

Processing. Besides whatever comments naturally come out of reviewing all the reports, it seems important in the processing to focus on how the students calculated an answer for an area of 1,000. Many students determine a pattern based on just the two examples (e.g., areas of 1 and 2)! Some students who generalize based on just the two examples get a correct answer, while some get an incorrect answer. In evaluating these two types of answers, I give the same grade to the correct and incorrect answer, since in both cases the students generalized with too little data. If this happens, it makes a good point of discussion for the processing. One method to emphasize this need in the discussion is to have the students compare the following solutions from three students in one class (for consistency, we will assume that each student found t = 2.71 for an area of 1 and t = 7.37 for an area of 2):

The first student

noticed that the difference between t for an area of 1 and t for an area of 2 is 4.66 which is about 4 2/3.... So in order to get the value of t for an area of 1000 you must multiply 999 times 4 2/3 and add it to 2.71. (Therefore,) the value of t for an area of 1000 is 2.71 + (4 2/3)(999).

The second student

found that when you square the value for the area equal to 1 the answer is very close to the value of t when the area equals 2. I (the student) therefore concluded that when you put the value of t (when the area equals 1) to the power of the area, you will find the corresponding the value for the area equal to 1,000 (t = 2.71^{1000}).

The third student

decided to take the square root of t when the area under the curve equals 2.... That number is surprisingly close to 2.71, therefore I assert that t equals 2.71 raised to a power equivalent to the desired area under the curve. I tried this formula with areas of a half and 0 and it seemed to hold. (Therefore, I think) t would equal 2.71^{1000} when the area under the curve equals 1000.

After student discussion, the teacher should guide the students to the conclusion that the first and second student made the same reasoning error: generalizing with too few examples; therefore, from a problem solving point of view the two students should receive the same grade *even though one student obtained the "correct" answer*. In contrast, the third student formed a hypothesis and tested it with two additional values.

Suggested Format. The students work individually on this task. The oral reports, if any, and the processing will require a portion of a class period. The project can be completed in five to ten school days.

Note: This area problem can be compared to the G(X) problem if you notice that the function A(t) = the area bounded by y = 0, y = 1/x, x = 1 and x = t has the property that A(ab) = A(a) + A(b).

Problems 3 and 4. "Transfinite Numbers, Part 1," and "Transfinite Numbers, Part 2." These two problems are difficult but are very well received by excellent mathematics students with some experience solving nonroutine problems. Here I will summarize the directions:

Transfinite Numbers, Part 1, Overview: The students are asked to compare the size of 14 sets including ten infinite sets. The restrictions of the problem force the students to discriminate between the size of the various infinite sets and to justify their answers.

Teaching suggestions: Give out the handout below and discuss. Be sure to make clear the implications of the stated restrictions; i.e., (a) the restriction that no more than five sets can be determined to be the same size implies that it is not acceptable to label all the infinite sets as being equal in size, and (b) the restriction that at least two groups of infinite sets should be the same size implies that it is not acceptable to label the infinite sets from smallest to largest with no infinite sets being of equal size. It should be noted that the purpose of these restrictions is to make the problem a "better" nonroutine problem by preventing them from making a trivial ranking based on non-important differences. Encourage the students to consider the hint; that is, if they can figure out why they are able to compare the size of the sets {1,2,3,4}, {a,b,c,d} and {1,2,3,4,5} easily, perhaps they can extend that reasoning to the infinite sets.

Solutions. On the one hand, I have never been satisfied with the solutions students have submitted for this assignment and, at times, I have questioned the value of this nonroutine problem as well as the second nonroutine problem on transfinite numbers, due to the difficulty of the problems. Yet when I asked students to evaluate the curriculum of nonroutine problems they worked on over the year, the two problems on infinity were clearly rated as the most difficult but also were rated as most interesting and the lessons from which the students felt they learned the most. In discussing these lessons with the students I felt convinced these two lessons are quite valuable, even though I have been disappointed in the quality of the students' solutions.

After the assignments are collected, you certainly want to discuss students' solutions. In addition, for enrichment and in preparation for the lesson, Transfinite Numbers, Part 2, you should discuss Cantor's Theory of Transfinite Numbers. Briefly, his major assumption was that he considered two sets to be of the same cardinality if at least one one-to-one correspondence between the two sets could be established. By this definition {1,2,3,4,...} and {2,4,6,8,...} are of equal cardinality because the 1–1 correspondence of n ↔ 2n can be established. Similarly, {−5,−4,−3,−2,..} and {1000,2000,3000,...} can be shown to be the same cardinality. By a more complicated proof, the set of all rational numbers can be shown to be in 1–1 correspondence with {1,2,3,4....}. One of Cantor's major accomplishments was to demonstrate that the set of all real numbers between 0 and 1 has a larger cardinality than the set {1,2,3,4,...}. He did this by assuming that they were the same size and reaching a contradiction. He argued that if they were the same cardinality there would be at least one 1–1 correspondence between the two sets which could be represented as follows:

$$1 \leftrightarrow .a_{11}a_{12}a_{13}a_{14}\cdots$$
$$2 \leftrightarrow .a_{21}a_{22}a_{23}a_{24}\cdots$$
$$3 \leftrightarrow .a_{31}a_{32}a_{33}a_{34}\cdots.$$
..
..
..
$$n \leftrightarrow .a_{n1}a_{n2}a_{n3}\ a_{n4}\cdots a_{nn\dots}$$
..
..
..

where a_{ij} is the jth digit in the ith number.

Cantor showed that there was at least one real number between 0 and 1 not on the list. He constructed such a number by choosing the nth digit of the decimal expansion to be any digit but a_{nn}. This number is not on the list because at least one digit of the number differs from each number of the list; e.g., the new number differs from the number corresponding to n because the nth digit is different. Contradiction. QED. Note: The actual proof is much more formal mathematically; for example, addressing the issue that there is more than one decimal expansion for some numbers (e.g., $1 = 1.0$ or $.999\dots$).

If we say that the number of elements in the set $\{1,2,3,4,\dots\}$ is $\aleph$, then the number C_0 = the number of real numbers between 0 and 1 can be thought of as equaling 2 raised to the $\aleph$ power (remember 2^n = the number of subsets of a set of n elements. C_0 can be thought of as the number of elements in the power set of $\{1,2,3,\dots\}$). Going back to the proof, one could say that we could have picked any of 9 digits for the nth digit in our new number, so the number of possible numbers we could have constructed could be named 9 raised to the $\aleph$ power.

By a proof beyond the scope of this paper, it can be shown that the number of elements in the set of all functions from (0,1) onto (0,1) is the third transfinite number. Also, using projections, it can be shown that the set of reals between 0 and 1, between −1 and 1, between 0 and 1,000 and between 0 and ∞ are all of equal size (C_o).

Transfinite numbers, part 1, Worksheet
Compare the size of the following sets:

$\{1,2,3,4\}$
$\{a,b,c,d\}$

{1,2,3,4,5}

{1,2,3,4...}

{2,4,6,8,...}

{−5,−4,−3,−2,...}

{the number of grains of sand in the Gobi Desert to a depth of 10 ft.}

{1000,2000,3000,4000,...}

{all functions from (0,1) onto (0,1)}

{all real numbers between 0 and 1}

{all real numbers between −1 and 1}

{all real numbers between 0 and 1000}

{all rational numbers between 0 and 1}

{all real numbers between 0 and ∞)}

Explain your rationale for labelling one set as being larger than another set. Your theory should not allow for any more than five sets being the same size. Also, at least two groups (two or more sets) of infinite sets should be the same size.

Hint: How did you compare the first three sets? Think about it! Can this method of comparison be extended?

Transfinite numbers, Part 2, Overview: The students are given ten expressions to simplify. The expressions involve one or more of the first three transfinite numbers.

Teaching suggestions: Give the students the handout below. To prepare the students for the assignment, tell them that you are going to review some concepts from the lesson Transfinite numbers, part 1 and some additional ideas that will help them simplify the ten expressions. You might suggest that they take notes on your comments. Review the following ideas from the lesson Transfinite numbers, part 1: (a) two sets have the same cardinality if at least one 1–1 correspondence can be established between the two sets; (b) the sets {1,2,3,4,...}, {−5,−4,−3, −2,...}, {1000, 2000, 3000,...} and {2,4,6,8,...} have the same cardinality which we represent by $\aleph$; (c) the sets of all real numbers in the intervals (0,1), (−1,1), (0,1000) and (0, ∞) all have the same cardinality which we represent by C_0; (d) the set of all functions from (0,1) onto (0,1) is the largest set discussed and the cardinality of this set is represented by C_1, (e) the number of subsets of a set of n elements is 2^n, (f) C_0 can be thought of as being equal to the number of subsets of a set with $\aleph$ elements and (g) 2^n increases much quicker than n^2 for n large.

Add the following hints: (a) point out that it can be shown that the number of lattice points (a,b) such that a and b are positive integers is equal to $\aleph$, (b) the probability of picking a rational number from the set of real numbers is zero, meaning that if you claim the probability is some small number epsilon you can show that the probability of the event is less than epsilon and (c) tell them the story of Hilbert's Hotel, a hotel with an infinite number of rooms numbered 1,2,3,4,…. in a line from left to right. One day all the rooms are occupied and a person asks for a room. The desk clerk replies that there is no problem and proceeds to announce to the guests, "Please move to the room to your right (i.e., $n \rightarrow n+1$)." Then he gives the new guest room 1. Soon an infinite group of new guests arrives asking for rooms. Again the desk clerk replies that there is no problem. He announces: "All guests move to the room whose number is twice your present room number (i.e., $n \rightarrow 2n$)." He then assigns the odd numbered rooms to the new guests.

It should be noted that number 5 on the worksheet, "simplify $\aleph - \aleph$," is the most difficult problem. This number is undefined because one can argue that it equals 0, or any finite number or $\aleph$ itself. Students are not generally aware of this difficulty so I like to alert them to the difficulty. Of course, another option is to not give them a hint and have a significant discussion of the problem after they turn in their solutions. If structured correctly, this challenge can add meat to the assignment.

Solutions: For each problem, I will give the solution according to Cantor's Theory and one example of an explanation. (1) $\aleph + \aleph = \aleph$: $\{1,3,5,7,…\}$ and $\{2,4,6,8,…\}$ both have $\aleph$ elements or together $\aleph + \aleph$, but the combined set $\{1,2,3,…\}$ has $\aleph$ elements. (2) $5\aleph = \aleph$: $\{1/5,2/5, 3/5,4/5,1,6/5…\}$ can represent $5\aleph$ elements, but there is a 1–1 correspondence between that set and $\{1,2,3,…\}$; i.e., $n \leftrightarrow n/5$. (3) $\aleph^2 = \aleph$: The number of lattice points can represent a set with $\aleph^2$ elements; recall the number of lattice points is equal to $\aleph$. (4) and (9) $C_o - \aleph$ and $C_o + \aleph = C_o$: You can argue that this is like removing a bucket of water from the ocean or use the fact that the probability of selecting a rational number from the reals is O. (5) $\aleph - \aleph$ is undefined; as previously remarked, this can be shown to be equal to 0, or any finite number, or $\aleph$. For example, these two sets have $\aleph$ elements: $\{1,2,3,…\}$ and $\{5,6,7,8,…\}$; $\aleph - \aleph$ can represent the number of elements in the set formed by removing the elements of the first set from the second set; that is, 5. (6) and (8) $\aleph - 2$ and $4\aleph + 1$ both are equal by comparing the following sets to $\{1,2,3,4,…\}$: $\{3,4,5,…\}$ and $\{0,1/4,2/4,3/4,1,5/4….\}$. (7) and (10): 2 raised to the $\aleph$ power $= C_o$, and $2^{Co} = C_1$: This can be argued by using the fact that 2^n = the number of subsets in a set of n elements.

Worksheet: Transfinite numbers, Part 2

Simplify the following:

(1) $\aleph + \aleph =$ (2) $5\aleph =$ (3) $\aleph^2 =$

(4) $C_o - \aleph =$ (5) $\aleph - \aleph =$ (6) $\aleph - 2 =$

(7) 2 raised to the $\aleph$ power $=$ (8) $4\aleph + 1 =$ (9) $C_o + \aleph =$

(10) $2^{Co} =$

Where $\aleph$ = number of elements in the set $\{1,2,3,4,\dots\}$, C_0 = number of real numbers between 0 and 1, and C_1 = the number of functions from $(0,1)$ onto $(0,1)$.

Provide documentation and support for your answers.

Problem 5. "Connecting Points": Students devise a procedure for connecting random points (at least 15) on a sheet of paper so as to minimize the distance required to connect them. The students are required to write a procedure that is clear enough that another person can follow and such that if two people follow the procedure the results would be the same. The procedure is tested by other students (or adults) on a sheet (8 1/2 × 11 paper) of random points (at least 15 points) provided by the teacher.

Student problem statement: The group task for this problem is to develop a procedure to connect random points on a sheet of paper so that the total distance is minimized. The purpose of the task is to give you practice solving a nonroutine problem which requires you to develop a written procedure. Specifically:

(1) Each group is required to write a procedure that is clear enough that another person can follow it and such that if two people follow the procedure correctly, then the results would be the same. The procedure will be tested by other students on a sheet of paper (8 1/2 × 11") of random points (at least 15) provided by the teacher.

(2) Each group will prepare a written description of the procedure for connecting the points. This description must be clear enough that another student not in your group can follow the procedure. Also, as stated above, if two students use the procedure on the same set of points their solutions should be the same. Your procedure will be tested by at least two other students. In addition, your group will prepare a description of how you developed your procedure, including any alternative methods you considered.

(3) Each group will prepare an oral presentation of two minutes or less on the gathered information. I will select the member of your group to present the oral report.

Teaching suggestions. You can introduce the problem by stating the following: The telephone company wants to develop a procedure that any (qualified) employee can follow which will tell the employee how to connect telephone poles using the least amount of wire. Emphasize that if any two employees follow the procedure correctly they should end up connecting the poles the same way (see Figure 11.2).

Put the following diagram on the chalkboard:

Ask the students what they believe the best way would be to connect these points (minimizing distance). In addition to the correct answer, student responses may include (see Figure 11.3):

However, the best answer is (see Figure 11.4):

The above example should alert the students to the difficulty of the problem which will involve at least 15 points. Remind the students that their assignment is to develop a procedure for connecting points (randomly placed by the teacher on an 8 1/2 × 11″ piece of paper) so as to minimize the distance.

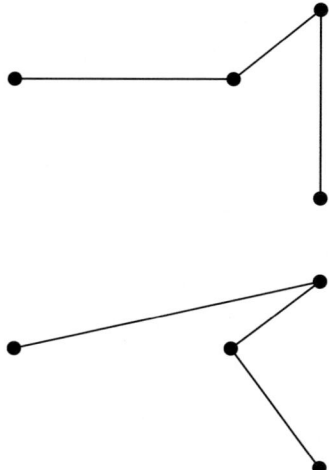

Figure 11.2 Four points, not connected.

Figure 11.3 Two sample student attempts to connect the four points.

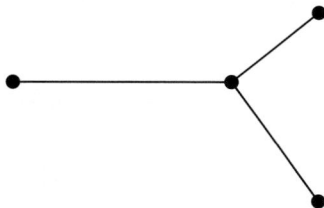

Figure 11.4 The best solution, connecting the four points.

Emphasize that any two students (or teachers) should be able to carry out the procedure and obtain the same result. Make sure the students understand that the teacher will provide some random points (at least 15) on a sheet of paper to be connected, and that their procedure will be tested on these points by comparing the total distance needed to connect the points with the distance required for other student procedures. Before the students finalize their procedures, you may want to have the groups exchange procedures, read the procedure given to them and identify any directions that are unclear, and then return the sheets with comments to the original group.

A sample page, "Connecting random points: Worksheet," with a set of random points is below. Students can use this sheet (or other sheets you provide or they create) to test their procedure.

After you have collected the papers and read them, the following activity can be fruitful: (a) give back the procedures and have the students exchange procedures, then read the procedure given to them and identify any directions which are unclear; (b) give back the procedures and questions to the original students (who developed the procedure) and have the students rewrite any unclear directions; (c) give each student a handout of random points and another student's procedure with the following assignment: follow the procedure and measure the total distance in centimeters and (d) discuss the results, perhaps the next day. An additional option is to have each procedure tested by two other students, thereby checking whether the procedure satisfies the condition of generating a unique result. This activity tends to focus the student's attention on the skill of developing algorithms. They get direct, concrete feedback on the clarity and effectiveness of their algorithm and are exposed to a variety of other solutions. Also, when the projects are complete, have the groups give their oral presentations and collect the written reports.

Solutions. Theoretically the best solution (assuming that line segments connect only given points; i.e., you cannot add points to the paper) is to first connect each point to the point nearest to it; this process results in a number of unconnected "links." Then connect links by connecting the two closest points of nearby links. Some students discover this method or equivalent methods, some of which take more time or include unnecessary steps. For example, one student first developed a method for numbering the dots (remember, two people using a method should get the same result) and then had the person "measure all the distances between all dots and record in a table for future reference." The process the person follows after these initial steps was equivalent to the best solution.

The following solution is from two students who worked together. Although the solution does not document initial methods, the account is entertaining and their solution is equivalent to the best solution:

Whilst working on a particularly tenacious connect-the-dots, it occurred to me that using the standard draw-lines-following-sequential-order-of-labeled-points technique is not necessarily the most efficient method of unifying the isolated spots. The possibility of another method even existing was so scintillatingly challenging that I vowed to define the operational parameters of a superior procedure or die, and set to work with a colleague of similar bent for grave research.

After long hours of studying random patterns of dots generated by bouncing a small wetly painted ball on a piece of cardboard (and carefully considering the potential philosophical ramifications of our actions), we developed a system for creating an optimized network of line segments. Our procedure is at once crude, yet elegant, awkward, but intuitive, of a technical nature akin to hyper-spatial mathematics, yet simple to implement, colorful, but not flashy, and requires the implementer to follow these steps: (1) Obtain 1 compass, 4 fluorescent color markers (possibly more), and a paper about which random dots have been strewn. (2) Pick a dot, any dot. Draw a line from it to its nearest neighbor with one of the markers, using the compass to delineate between candidates. If there are two (or more) dots equidistant, do not draw a line. (3) Select another dot and repeat procedure two, using the same colored marker. Do this until every dot has a line drawn to, or from, it. (4) Change fluorescent markers. You have just established the initial sub-networks of the master network. To continue, you must now connect these sub-nets, in a similar manner, to create the next network level. Starting with one sub-net, determine which dot of that sub-net is nearest to what dot of another sub-net, connect these two. (5) Proceed with step four until each sub-net is connected to another sub-net. You will then have constructed a further network level and must then repeat step four, treating the newly constructed sub-nets into further sub-nets until the entire network has been established. (6) Pat yourself on the back.

It should be noted that the above solution is not typical (one of the two students went on to MIT); it is unusual that students discover the best solution. However, it does give a sense of the fun students can have developing algorithms in the context of solving a nonroutine problem. The next solution demonstrates a student's procedure that is typical in quality to the type of procedure many students develop:

(1) Create a direction vane in the upper right hand corner of the paper such as this (see Figure 11.5):

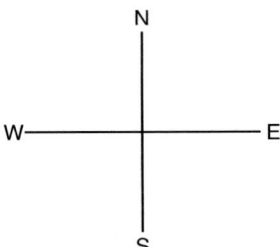

Figure 11.5 A direction vane labelled N, W, S, and E.

(2) Next, compare the distance between the northern and southern most points to the distance between the eastern and western most points. Determine which distance is greater.

(3) Now, if the smaller distance is equal to or less than 1/4 of the larger distance, then go to Procedure B; otherwise, go to Procedure A.

Procedure A:

(1) If the larger distance is from west to east, divide the points in half making the dividing line from west to east. Or else, if the longer distance is from north to south, divide the points in half making the dividing line from north to south.

(2) Next, turn the paper so the line is parallel to a west-east line.

(3) Go above the line to the left most point and number it one.

(4) Continue this from left to right staying above the dividing line.

(5) When you have finished above the line, go below the line to the right most point and continue moving from right to left, now.

(6) Connect the dots following the numbers.

Procedure B:

(1) Turn the paper so the longest distance is parallel with you.

(2) Start at the left and number the points from left to right.

(3) Connect the points following the numbers.

In documenting this procedure, the student commented:

When creating this method, I felt cutting down on the sideways motion as much as possible was extremely important. In order to do this, I divided the shorter width into two sections and then connected the dots on each side of the line. An exception to this is when the points are in a straight line or close to a straight line. That is why I've created procedure B.

Processing. In the processing of this project, focus on how groups dealt with the problem of creating directions which others could follow and resulted in a unique solution. A discussion of what type of feedback groups got when they exchanged procedures might be fruitful.

Suggested Format. The students work in groups of three and are given a portion of two periods to develop their procedure and prepare their oral report. You will need a portion of one class for groups to exchange procedures and try them out. The oral reports and the processing will each require a portion of a class period. The project can be completed in seven to ten school days. This project can be completed individually rather than in groups.

Enrichment and Extension. An unsolved problem is to determine a general procedure for constructing the shortest closed path for connecting random points, allowing for adding points. There is a theorem that the shortest path cannot include two lines that cross. Students could try to develop procedures for this type of path and compare their results on two or three sample sets of points.

Sample random practice points (see Figure 11.6):

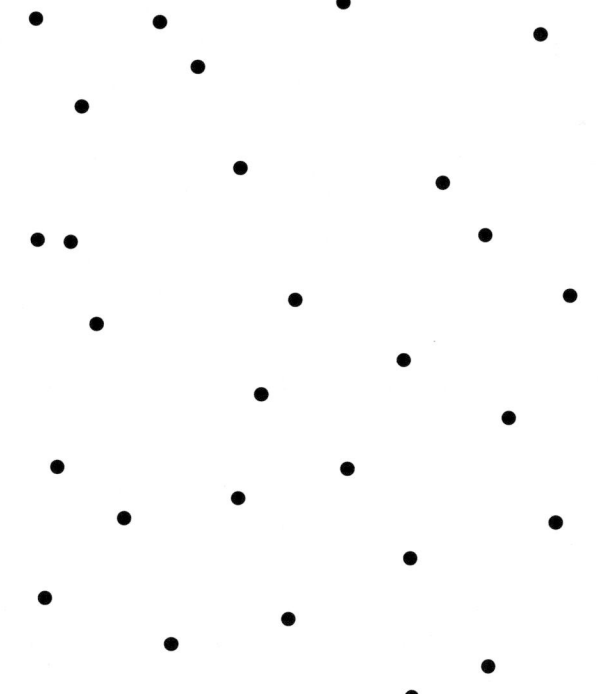

Figure 11.6 Sample page of random points, not connected.

Additional nonroutine problems not field-tested

Four potential nonroutine problems I have not field-tested are:

(1) "Miles of Road." Students determine the number of miles of road in their town. As long as there is not a written answer in the town's material, this could be an interesting problem. If your school is located in an urban neighborhood, perhaps a section of the neighborhood would be appropriate.

(2) "Hawk Migration." Students are given data concerning the hawk migration over Hawk Mountain, Pennsylvania or Cape May, New Jersey in one year and are asked to predict the migration in a second year given weather conditions and total number of migrants. Data is available from both locations. Of course, any migration location that keeps yearly records would be appropriate.

(3) "Tree (weight and height)." Determine the weight and height of a big tree. This problem would be most appropriate before students studied trigonometry.

(4) "$5 project." This nonroutine problem was implemented by a teacher, Mark Witvliet, a student in the MA program in Holistic and Integrative Education that I directed at California State University, San Bernardino as part of an assignment in the program. He reported that he borrowed the idea from a teacher in Iowa. He gave each of his 53 eighth grade students $5 with the assignment to grow the money during the next five weeks. He required three progress reports during the five weeks plus one final report. He had a number of rules including prices needed to be fair, no investing, no selling at school, students could not take out loans and limited help from parents. At the end of the project, any money in addition to the $5 would go to a charity of the class's choice. He answered students questions and certainly provided necessary scaffolding when needed based on the progress reports.

Results: In addition to the initial $265 returned to Mark, the students raised an additional $670 distributed to two charities! In total, 98% of the students turned in at least $5, 75% at least doubled their money and $17.63 was the average amount turned in. The ideas of the students included a student that (a) bought a bucket, soap and towels and washed cars, (b) bought stencils and paint and painted house numbers on the curb and (c) bought ingredients and made dog biscuits. Mark commented in an email the next year that he did it again with a few changes and raised $3,800.

Additional problems can be found in professional journals for teachers. For example, here are examples from three different journals. First, perhaps the best source is the National Council of Teachers of Mathematics (NCTM) website and journals. Three examples from NCTM are problems for designing a house (Boehm, 1998), designing a city park (Tepper, 1999) and designing a garden (Greeley, 1998). A second good source in selective issues is Educational Leadership, the journal probably most referred to by principals and superintendents. Each issue addresses a topic considered important with a variety of short articles on the topic by leaders in the field including topics such as place-based education, the maker movement and problem based education. For example, one article (Greene, 2017) discusses the project of "engineering" a better tent. Third, the journal, Rethinking education, focuses on social justice issues such as a project to investigate sexism in toys in toy stores (Hoffman, 2005–6). These are just a few projects discussed in those journals, as well as potential additional nonroutine problems in additional journals and written resources. My one caution is to look for problems that do not need to be connected to specific content objectives. Remember the principle of pedagogy for nonroutine problems that problems without significant content objectives or prerequisite skills are more likely to provide a better focus on improving their ability to solve nonroutine problems, especially in the beginning of the curriculum.

Introductory unit

As mentioned in Chapter 2, the essence of the curriculum is that the students develop an understanding of the three steps of solving nonroutine problems and effectively use the three steps to solve meaningful nonroutine problems in their life. The key instructional technique is to have students solve nonroutine problems at the appropriate level of difficulty in a variety of settings and through the processing of the problems help students gradually improve their skill solving nonroutine problems. The first few years I worked with average students on learning how to solve nonroutine problems I did not include an introductory unit on the three steps. My inclination was to just have the students work directly on nonroutine problems and process what happened. It became clear that most students had much difficulty making the transition from never attempting problems of this nature, to jumping right in and solving nonroutine problems, they needed some type of bridge to work productively solving nonroutine problems. Consequently, I developed and field-tested a unit that introduces students to the significance of the three steps of solving nonroutine problems and gives them a sense of how the steps work. Specifically, in this unit, the students are introduced to the definition of a nonroutine problem, given an explanation of the three steps in solving a nonroutine problem, given a variety of examples of how the steps work, asked to generate additional real-life examples and tested on their understanding of the three steps. One approach to this unit that worked well for me was the following sequence:

(1) Introduce the students to the definition of a nonroutine problem and have a conversation with them concerning the Census Taker problem. As discussed in Chapter 2, I try to help students understand how their attempts to solving the problem are both similar (initially) and different (trying something and persistence) to attempts by mathematicians. This conversation has been effective starting the process of understanding how to solve nonroutine problems.

(2) Review at least three examples with the class from real life that illustrate the three steps. Four that I like are the work of Charles Darwin in discovering that some plants eat animals and why (Brown, 1988); Dr. Alex Shigo's work "to find out why trees rot – or, more to the point, why they often don't" (Chapline, 1986); Barbara McClintock's work concerning the genetic organization of corn (Keller, 2003) and the jacket notes on Paul Simon's album Graceland (1986) that had a significant effect in increasing the influence of African music on American popular music.

(3) Have the students individually or in small groups research additional examples. Good sources of examples abound in the fields of science, music and art. For example, the book "Journeys of women in science and engineering" (Ambrose et.al., 1997) provides many examples of women in science.

(4) I assess the students' understanding of the steps by giving a test for which the student picks three examples to study for the test for which she/he will explain how that person applied the three steps. I have provided students with some sample examples with documentation of the three steps that they can pick for the test if they wish; I am only assessing a basic understanding of the three steps based on examples discussed in class.

In my experience, these four steps are successful in giving most students the basic understanding of nonroutine problems to start the curriculum. There are certainly other ways to construct this unit, what is important is that you provide some instruction before they begin work on actual nonroutine problems.

Conclusion

As mentioned in Chapter 10, I have tried in this book to provide clearly all the information and guidelines you need to implement a sequence of nonroutine problems; however, I know that actual implementation *in your context* can be more complicated than the guidance given here. Therefore, I encourage you to contact me if you have a question concerning implementation at rlondon@csusb.edu.

References

Ambrose, S., Dunkle, K., Lazarus, B., Nair, I. and Harkus, D. (1997). *Journeys of women in science and engineering.* Philadelphia, PA: Temple University Press.

Boehm, E. (1998). House plans. *Mathematics Teaching in the Middle School*, 3(4), 297–9.

Brown, L. (1988). Predator Plants. *Connecticut Audubon*, Summer, 1988, 29–32.

Chapline, J. (1986). A new testament of tree care. *Country Journal*, June 1986, 66–74.

Greeley, N. and Offerman, T. (1998). Now and then: Garden designer. *Mathematics Teaching in the Middle School*, 3(6), 428–33.

Greene, K. (2017). Problem solving in practice: Engineering a better tent. *Educational Leadership*, October 2017, 75(2), 44–5.

Hoffman, S. (2005–6). Miles of aisles of sexism: Helping students investigate toy stores. *Rethinking Schools*, Winter 2005–6, 20(2), 44–7.

Keller, E. (2003). *A feeling for the organism: The life and work of Barbara McClintock.* New York: Holt.

Tepper, A. (1999). Designing a city park. *Teaching Children Mathematics* 5(6), 348–52.

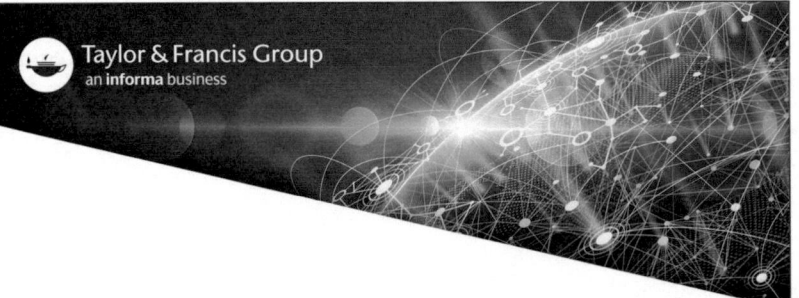